THE ELECTRONIC
CONFESSIONAL

By Howard R. Lewis

With Every Breath You Take
Growth Games (with Dr. Harold S. Streitfeld)
RN's Sex Q & A: Candid Advice for You and Your Patients
 (with Armando DeMoya, M.D., and Dorothy DeMoya, R.N.,
 M.S.N.)

By Howard R. and Martha E. Lewis

Psychosomatics
The Medical Offenders
The Parent's Guide to Teenage Sex and Pregnancy
Sex and Health (with Armando DeMoya, M.D., and Dorothy
 DeMoya, R.N., M.S.N.)
Sex Education Begins at Home
The People's Medical Manual: Everything You Need to Know
 About Health and Safety

THE ELECTRONIC CONFESSIONAL

A SEX BOOK OF THE 80'S

HOWARD R. AND MARTHA E. LEWIS
Edited by Jean Arbeiter

M. Evans and Company, Inc.
New York

Library of Congress Cataloging-in-Publication Data

Lewis, Howard R.
  The electronic confessional.

  Includes index.
  1. Sex—Miscellanea.   2. HSX (Videotex system)
3. Sex—Information services.   I. Lewis, Martha E.
II. Arbeiter, Jean S.   III. Title.   IV. Title: Sex book
of the 80's.   [DNLM: 1. Counseling.   2. Information
Systems.   3. Sex.   4. Sex Behavior.   WM 55 L673e]
HQ31.L46   1986   306.7   86-8928

ISBN 0-87131-478-9

M. Evans and Company, Inc.
216 East 49 Street
New York, New York 10017

Design by Lauren Dong

Manufactured in the United States of America

9 8 7 6 5 4 3 2 1

FOR VIDEOTEX PIONEERS
HSX members
HSX section leaders
CompuServe colleagues

This book does not substitute for the medical advice and supervision of your personal physician. No medical therapy should be undertaken except under the direction of a physician.

Contents

The Electronic Confessional: Introducing the Human Sexuality Computer Service

Walter, (31): I'm very lonely. My job and where I live make it hard for me to meet other people.

Felice, (28): I don't want to leave my husband. But I long to talk with a man who shares my interests and understands me.

Herb, (33): Am I normal? I'm in my thirties and have a wife. Still and all, I masturbate.

Margaret, (24): I've been too involved in my career to ever have intercourse. Before I finally get into bed with a man, I want to be able to express my sexual needs.

Kevin, (17): I'm a high school junior, and I know I'm gay. I have a million questions—and no one I can ask.

Ilka, (29): I must be an unnatural mother because I wish I'd never had my baby. I'm desperate to talk about this, but I'm too ashamed.

Ed, (47): My wife and I find sex blah. How do other long-marrieds spice up their sex lives?

Harriet, (41): Bob and I are intrigued by the idea of having sex with other married couples. But first we want to talk to swingers to find out how swinging affects their marriages.

Roy, (52): I'm recovering from a heart attack. How can I resume sex without getting another coronary?

Dorothy, (37): I walked in on my 15-year-old daughter having intercourse with her boyfriend. What do I do now?

Ian, (34): I'm terrified of getting AIDS and herpes. How can a single like me have sex safely?

Andrea, (29): I'm considering becoming a surrogate mother, bearing a stranger's child for a fee. I'm concerned about my medical and legal risks.

Michael, (38): The world sees me as the ideal husband, father and professional. No one knows my greatest pleasure comes from wearing women's clothes and makeup. I dare not reveal my transvestism, but I urgently wish to talk with other men like myself.

Jeff, (30): I think I've O.D.'d on sex. Ever since I was in college I've had all the sex I could want. Suddenly, this whole way of life has gone sour, and I haven't wanted sex for months. Is this weird? Dangerous to my health?

Human Sexuality

All these people found the human contact they desired, the information they needed, the advice they sought. They secured friendship. They gained support.

And they did so without leaving their homes—by means of a computer service we named Human Sexuality.

Human Sexuality (abbreviated HSX) is part electronic magazine, part electronic meeting ground. Imagine picking up a copy of *Time* or *Newsweek* or *Reader's Digest* and being able to read not only the current issue but any article you desire from any previous issue. Also imagine being able to talk to any other person who's reading the magazine at the same time. Imagine joining conversations that have been going on between readers for days, adding your comments for others to respond to when they next pick up the magazine. And imagine your remarks becoming a permanent part of the magazine.

So it is with HSX. By typing "commands" (codes) at your computer keyboard, you can read over a thousand articles written for the general reader about emotions, relationships, and sexual functioning and physiology. You also can converse with other people who are using part of the service. HSX is both a new form of journalism and a new way for people to relate.

The Videotex Medium

HSX is a product of "videotex." The *video* screen of your home computer displays *text* from a distant computer you've dialed through your phone line. HSX is part of a rapidly expanding development in communications: the marriage of the computer to the telephone. Thanks to telecomputing, you have at your fingertips a rapidly expanding ''on-line'' world.

Already, about 1.5 million Americans subscribe to videotex services. The number of new subscribers has been growing by more than 20 percent a year. Most seek information. They tend to regard the microcomputer as a gateway to a vast electronic library, and are willing to pay hourly fees and surcharges for the speed and convenience.

Assume you use videotex to the hilt, as some subscribers do. To keep up with the news, you display the latest items from the Associated Press or United Press International on your computer screen. For the latest stock prices and financial news, you look at *Quotron*, *Dow Jones* and other investment services. To keep abreast of your field, you ''log in'' to *NewsNet*, which offers the contents of over 300 newsletters. Your interests are on file, so the computer shows you only pertinent items.

Journals? Magazines? Newspapers? You read complete articles on your computer screen, running the gamut from technical literature to the *Baton Rouge Morning Advocate*. To research a topic, you consult bibliographies such as *Dialog*'s 200 indexes with over 75 million listings.

If you're a lawyer, you find precedents with *Lexis* or *Westlaw*, much easier than poring through law libraries. If you're in medicine, you consult *Colleague* for relevant chapters from texts. If you sell real estate, you draw on a videotex multiple listing service. If you travel, you plan your flights with the electronic edition of the *Official Airlines Guide*. If you're asked for credit, you regard videotex credit-reporting services like *DunSprint* and *TRW* as extensions of your phone.

HSX on CompuServe

HSX is available through CompuServe, the largest of the videotex networks. CompuServe began in 1969 as the data-processing wing of an insurance company. It expanded into renting computer time to outside companies. In 1975 it was spun off as a separate ''time sharing'' service.

To make use of idle capacity after the workday rush was over, the firm in 1979 made its computers available to hobbyists. For a small hourly charge, programmers with kiddie-car-sized machines could drive a Sherman tank. The idea caught on, and the company added more and more features.

Now CompuServe is a utility with about eight hundred services and three hundred thousand subscribers. Grolier's *American Academic Encyclopedia* is on line. So is The College Board, which helps youngsters prepare for standard achievement tests, choose a college, and get financial aid.

In the Electronic Mall you can browse through some seventy-five shops, seeing the merchandise through extended descriptions and ordering directly on line. Site II, which analyzes demographic data, is one of many statistical programs. Several banks make your computer screen a teller's window. You can play blackjack, backgammon, and dozens of other games—including MegaWars III, a multiplayer space fantasy so complex that players have been immersed in it for days.

Reaching HSX

To reach HSX (or any other videotex service), you need a device called a modem to link your computer to a phone line. The modem gets its name from *mo*dulator–*dem*odulator—it modulates (modifies) impulses from the computer so they can be recognized by telephone equipment, and demodulates signals received from the other end so your computer can understand them. Lower-priced modems transmit about 300 words a minute, higher-priced ones about 1,200.

The number you dial ties you into a special telephone network, one set up to carry not human voices but computer signals. Such networks—the biggest ones are Tymnet and Telenet in the United States and Datapac in Canada—have phone numbers ("nodes") in most major cities. CompuServe supplements these with nodes of its own. Thus, even though you want to connect with a computer thousands of miles away, you may make only a local call.

When you finish dialing, you hear a ring or two. Then the network's computer answers, emitting a high-pitched tone. Make the right moves with your modem, and this whistle translates onto your screen as a request for whom you wish to call. If you're using

Tymnet, you're told, "Please log in." To indicate the CompuServe Information Service, you type: "CIS."

Like the address on a letter, "CIS" carries you from one connecting point to another. The impulses triggered by your keystrokes may travel by wire, relay station, and satellite. In an instant, they reach Columbus, Ohio, entering one of CompuServe's thirty-one Digital Electronic Corporation mainframes.

You're asked for your ID number and password. Thereafter, to reach Human Sexuality, you type the command: "GO HSX."

"GO" reflects the astonishing sensation many users get of *traveling*. Reading videotex has much in common with listening to radio: the words create pictures in your mind. A mental landscape forms, which you can navigate at will. CompuServe subscribers speak of "going over" to HSX, "dropping by" the Electronic Mall, "diving into" MegaWars. You never leave your chair. But day or night, spinning through the system in your mental spaceship, you're likely to find something interesting going on.

GOing to HSX

The first HSX "page" (screenful) you come to is intended to warn off those who might be offended. It cautions readers:

We put up the first edition of HSX in July 1983, drawing on our backgrounds as medical writers and editors. Our previous four books had dealt with aspects of sexuality; we were editors of the journal *Sexuality and Disability* and were in charge of the "Sex Q&A"

column for the nursing magazine *RN*. CompuServe at that time was going through a period of explosive growth and was looking for new services. At first we thought of HSX only as a freelance project that would help us pay for the microcomputer we'd surprised ourselves by buying. We soon found we were inventing a medium.

HSX features findings from the new branch of health care called sexual medicine. This specialty draws on advances in urology, gynecology, psychiatry, pharmacology, endocrinology, and many other fields. Through this electronic magazine we seek to provide a helpful, authoritative source of sex-related information and advice.

Following the practice of medical journals, we brought together a distinguished group of consulting editors (listed at the back of this book). In our Statement of Purpose we told our readers, "Through Human Sexuality your computer provides you with a direct line to foremost authorities in sexual medicine. Our Consulting Editors and other specialists offer you knowledge and guidance on virtually any sex-related concern."

We put up editions more or less weekly. Videotex is an intensely personal medium—you talk to just the one person who's reading the screen. To emphasize our personal touch, the front page announcing the current highlights is headed:

After seeing what's new in this issue, you come to a "menu" (list of choices) that looks like this:

```
          HUMAN SEXUALITY MAIN MENU
: : : : : : : : : : : : : : : : : : : : : : : : : : :
   1 Special Features
   2 On-Line Transcripts
   3 Answering Your Questions
   4 Personals * * * Letters
   5 SUPPORT GROUPS A & B (FORUMS)
   6 Talk About Relationships
   7 * * Interactive Programs * *
   8 Welcome to HUMAN SEXUALITY
   9 INDEX: Find Your Subject
   10 Hotline: YOUR Messages
   |_
```

The "_" is the "cursor" (pointer) at the "!" prompt, where you type your choice.

In the Database

Most of these menu choices take you to our database, computer files set up for easy access. The conventional database is pretty dry reading. However, we've written for such magazines as *Family Circle*, *Consumer Reports* and *Reader's Digest*, and apply the informal, how-to style of those publications to discussions of sexuality. In videotex, the lines scroll down the screen faster than ordinary reading speed. To ease comprehension, we use very short sentences, with paragraphs seldom more than three lines long and lots of subheads to provide frequent breaks.

If you press 1, you come to our Special Features. These deal at length with large issues: "Arguing Over How Much Sex," "Answers About Masturbation," "How to Handle Sex Talks." Each Special Feature is broken down into sections that readers may select from a menu. One popular feature, for example, has this menu:

```
:: WOMEN AND ORGASM
:: Female Sexual Satisfaction:
   What You Need to Know and Do

1 What Causes Difficulty
2 The Clitoris and Orgasm
3 The G Spot
4 Achieving Multiple Orgasm
5 Hints for Sexual Exploration
6 Understanding "Frigidity"
7 Step-by-Step Suggestions
8 Vibrators and Orgasm
```

We also have, with our Special Features, manuals such as ''The Birth Control Guide'' and ''The VD Guide.''

''On-Line Transcripts'' (2) are conversations with experts. Psychiatrist Harold Lief warns about hazards for young marrieds. Venereologist Nicholas Fiumara discusses the emotional and physical aspects of herpes. Family physician Randall Rissman takes a young man through a testicular self-examination. Sex therapists Armando and Dorothy DeMoya offer a program for relieving premature ejaculation. Psychologist Thomas McDonald gives advice on overcoming computer addiction.

Our best-read offering is probably 3, ''Answering Your Questions,'' relatively brief replies to inquiries from readers. A sample menu:

```
:: ANSWERING YOUR QUESTIONS #22
 1 "I've Overdosed on Sex"
 2 "What Could Cause Impotence?"
 3 Finds Son & Friend Having Sex
 4 "Can Tampon Get Lost Inside?"
 5 "Erection Points Off Center"
 6 Sex After Spinal Injury
 7 When Hymen Needs Surgery
 8 "Don't Like Breasts Touched"
 9 Confidential to...
10 What's YOUR Question?
```

''Personals and Letters'' (4) covers a wide range of tastes and topics. In addition, we have ''Talk About Relationships'' (6). Can men and women be friends? How can one ease the pain of a breakup? What's the best way for a gay to come out? Readers share their experiences.

''Welcome to HUMAN SEXUALITY'' (8) is our masthead. In this menu choice are our Statement of Purpose and brief write-ups about us and our consulting editors. We also talk about HSX in ''Chat With Us.''

Print magazines get thrown away, and the articles in them disappear. HSX, however, is ever-expanding. Editions are cumulative, updated as needed but always available. Ergo, choice 9: ''INDEX: Find Your Subject.''

Interactives

''Interactive Programs'' (7) add a human dimension to computer-assisted instruction. Our Interactives simulate discussions between the reader and us. For example, this from ''Enrich Your Sexual Wordpower'':

When we score your answer we add information about sexuality and behavior. Say your name is Chris, and you reply ''3.'' An exchange like this follows.

The correct answer, incidentally, is ''2.'' In the same way, you can also take a variety of personality tests.

Hotline

Our readers can connect with us via our ''Hotline.'' If you choose 10 on the main menu you get taken to a special feedback file. You're asked for your first name and age, then your comment or question. These are stored for us until we retrieve them.

We've scattered menu choices for the Hotline throughout HSX, and on every ''Answering Your Questions'' menu. We get about 500 Hotline messages a week, the source for most of our questions, letters, and ''Talk About Relationships'' responses.

HSX's appeal derives in large part from videotex's unique capacity for combining intimacy and anonymity: You are alone with your computer. And while you're on-line with HSX, you're known only by whatever name you choose to give.

Such confidentiality invites remarkable candor. Our readers find they have little reason to hold back on their questions or refuse to share their experiences. As a result, we've become privy to the most intimate concerns of the people who use our service.

Support Groups

By pressing 5 on the main menu, you come to the Human Sexuality Support Groups. These are 500,000 self-help groups in the United States. In these groups, members share feelings and experiences and provide one another with assistance and encouragement.

The groups can't substitute for psychotherapy—but they can help members develop friendships, grow emotionally, and solve many personal problems.

For HSX, we created the videotex equivalent of support groups by adapting CompuServe's software for "forums" (also widely called special interest groups or SIGs). To visualize the HSX Support Groups, imagine a building somewhat like a conference center or college student union. There are many suites ("sections" or "subtopics"). In each section are electronic meeting rooms.

Let's go into the section "For Women Only"—our oldest support group, modeled after women's consciousness-raising groups. Most of our sections are open to all members, but "For Women Only" is closed. You're specially "enabled" (admitted) only if we feel assured that you're female.

One type of meeting room in the section is comparable to what you experience in everyday life: You and others meet in "real time," simultaneously sitting at your computers and "talking" to each other via your keyboards. You may chat in an open lounge— a "conference area"—where anyone can join in. Or you can talk privately, in small groups of two or more, where others are excluded. Your messages appear on each other's screens in transcript format. (For edited examples, see "Electronic Conversation.")

Videotex also transcends time and space. HSX offers a greatly expanded version of the hundreds of bulletin board services run by microcomputer hobbyists around the country. The mainframe can hold a large number of messages. It can also link your remarks to others by *subject*, irrespective of when or where you participate.

Thus, in a kind of conversation unique to videotex, minds can meet through an electronic "message board." Say you're in New York and you post a comment or problem. A minute later, Alice in Seattle replies to you. An hour after that, Leslie in Miami makes an observation to Alice. Next day, Terry in Chicago offers a suggestion to you, disagrees with Alice, and asks a question of Leslie. During the week, Jill in Dallas recounts a related experience she's had, and still more people join in the discussion.

By typing a command you can sort out the "thread" of this conversation from others that are taking place on the board. Although participants may be anywhere and join in at any time, the dialogue on the screen resembles the kind of back-and-forth you might hear in a living room. These remarks may be stored on an oxide-coated disk in Columbus, but—written in light—they come streaming across your screen with surprising immediacy.

Each HSX support group has at least one "section leader," a bright, articulate person who serves as a combination editor, columnist, and discussion facilitator. A 40-year-old executive we'll call Jan has been one of the leaders of "For Women Only." She lives on the West Coast with her daughter. She describes herself: "I was divorced a year and a half ago after a 'less-than-happy' eighteen-year marriage. I own and operate a large escrow company, which succeeds in occupying at least ten hours of each day. High pressure is very familiar to me!"

Jan's view of this on-line support group? "This is a special place for, by, and about women and the issues that affect us. Sort of our 'private corner,' where we can openly discuss our questions, problems, happiness, and whatever else is on our minds."

As a section leader, Jan's involved in a new journalistic form: creating and following through on a weekly "seeder" (discussion topic). The software limits the titles of seeders to 24 characters. Among Jan's have been "Your erogenous zones?," "Who wears the pants?," "Ever fake an orgasm?," " 'He's just TOO kinky!' "

All told, we have twenty groups, divided by the available software into two parallel forums, A and B. At this writing, in addition to "For Women Only," we have its counterpart, "Man to Man." "Dear Diane" offers advice from a psychotherapist. "Person to Person" doubles as a shyness workshop and place for posting personal ads. "Couples/Parents" deals with family relationships, including cohabiting partners.

"Teens Talk!" concerns problems of adolescents. "Gay Alliance" serves male and female homosexuals. "The Singles Club" is for unmarrieds. "Matters of Morals" deals with issues of religion and ethics. "Alternatives" is for people who engage in transvestism, fetishes, and other options. "Gay Youth" seeks to relieve the isolation common to young gays.

"Rusty's Pub" is a friendly café where you can discuss current sex-related news. In "Cope With Crises," a trained therapist offers help with problem situations. "Debbie's Diary" is one woman's highly personal journal. "Formerly Married" is for people who are divorced, separated, or widowed.

"Encounter Group" delves deeply into members' feelings. "Sex Ed for Adults" features in-depth seminars led by a medical school faculty member. "You and Your Body" deals with such issues as weight control, body image, and handicaps. In "Fantasyland," you can explore dreams, fantasies, astrology, the occult, and other phenomena of the unconscious. "Slug It Out!" is our verbal gym, where you can release your aggressions and hostilities.

Each section also has its own archive, a "data library" where you can read previous threads as well as articles, essays, poems, stories, and other miscellany submitted by members. The library also stores computer programs donated by members, which you can "download" (copy) free.

In addition, the forum at large has a member interest file, an electronic "personals" column. To find like-minded people, you can search the file for interests you share.

On-line Relationships

The medium tends to attract good writers whose thoughts flow unself-consciously through their fingers. In the HSX conference area, we asked a member a question. He thought a moment, typing "Ummm" as he pondered.

Personalities emerge. Alice comes over as cool and intelligent; her precise usage, her balanced phrases, her meticulous punctuation all suggest a certain starchiness. Leslie, with lavish use of capitals and exclamation points, seems MUCH MORE excitable!! Terry, a persistent misspeller, looks rather casual. Jill . . . a set of dots here . . . another set there . . . appears a bit . . . disconnected.

Participants in on-line discussions often laugh: "Hehehe." With their less-than ("<") and more-than (">") keys, they interject facial expressions ("<grin>," "<smile>"). They demonstrate affection ("<hug>"). They even add sound effects ("<clap> <clap> <clap>").

Although members may never meet or even speak on the phone, on-line relationships can have an enormous impact on their lives. In former years, letter-writers developed friendships and romances through their correspondence. Videotex combines this with the speed of ultrafuturistic technology.

People chat and argue. They fall in love, become jealous, break up, have feuds. Several marriages have resulted from on-line meetings. Some people engage in compusex, sharing interactive fantasies and feeling safe from disease and pregnancy.

One need never be lonely with HSX available, however great the physical isolation. There are always messages to respond to, people to talk with. Indeed, many people become addicted to computer relationships.

What's it like to connect with other people whose words are wafting across your computer screen? Observes an HSX member, *Ms.* magazine contributing editor Lindsy Van Gelder: "It's dizzyingly egalitarian, since the most important thing about oneself isn't age, appearance, career success, health, race, gender, sexual preference, accent, or any of the other categories by which we normally judge each other, but one's *mind.*"

We see many people in technical fields who are good at their jobs but have few social opportunities. On the computer they can meet people with their minds. They don't have to feel uncomfortable as they might in a bar. If a relationship becomes too threatening, they can simply stop typing. On the other hand, they have support from other people if they need it.

Mind-to-mind encounters presume that the people at both keyboards are committed to getting past labels and into some new, truer way of relating. Notes Lindsy: "Thoughts and emotions are the coin of this realm, and people tend to share them sooner than they would in 'real life.' "

Compassionate Community

One reason the HSX Support Groups are successful is that we enforce a code of conduct. You're likely to get the most out of any support group if you're speaking from your heart and welcome feedback. To help you feel free to do this, we've drafted an Agreement binding on all members and calling for honesty, respect, and confidentiality.

Thanks in large part to the Agreement, members openly discuss their thoughts. Computers may seem cold and mechanical, but they can foster a warm, supportive environment in which you feel safe in sharing your feelings, experiences and relationships. Your privacy and acceptance are assured. You can feel safe in candidly discussing personal matters. You can find sympathetic, intelligent people to connect with.

Members of the Human Sexuality Support Groups have agreed

to be sympathetic and serious about one another's concerns. By sharing in this commitment, they've created a compassionate community.

Sex in the 1980s

What does HSX tell us about what's going on sexually in America in the eighties?

The questions people ask the service reflect expectations that were hardly imaginable a few decades ago. Today people consider good sex to be something worth pursuing through information, much as they pursue good health through exercise. There's a feeling that satisfying sex is one of life's entitlements, something within everyone's grasp.

HSX readers want the latest medical information. The questions show a big emphasis on the practical. People are demanding scientific information because there has been an unexpected side effect of the sexual revolution: the vast increase in sexually transmitted disease and other health problems associated with sex. Increasing medical technology, in such areas as contraception and obstetrics and gynecology, has also complicated America's sexual life.

Another aspect of sex-as-entitlement is that people are refusing to relinquish sex because of illness, incapacity, or aging. We get lots of questions about how to have sex after bypass surgery, with multiple sclerosis, or with spinal-cord injury. Questioners want to know how to get around physical problems. They won't give up on sex because of various disabilities, as people of previous generations might have.

The people we're making close contact with—the HSX readers and members who share with us their sex-related concerns—are upper-middle-class Americans you might well know. What's going on in the hidden lives of these people next door? From the vantage point of our electronic confessional, we've learned that the ideal presented by the sexualized media is a far cry from the reality experienced by our readers.

The media hype about the sexual revolution initially led us to expect that all the young professionals who use this service would be sexually savvy. True, there is a great deal of sophistication expressed in many readers' queries. Nonetheless, many other questions reveal that the old sexual bugaboos are still with us. Many people are ignorant about basic anatomy. Others betray misinformation and myths.

A basic illustration of the gap between media hype and reality: The erotic literature, for all the words it expends on sexual function, deals virtually not at all with the most universal sexual outlet: masturbation. When we began HSX, we got more questions about masturbation than about the next three subjects combined. In as many ways as we could, we assured readers of the utter normalcy of masturbation. It's a source of some pride to us that we've evidently been successful. The worried questions have dropped off, and we're increasingly being asked how to do it better.

In the media, sex is portrayed as something that doesn't take any work—something that males especially should know how to do well, as if the knowledge came with their genes. From our vantage point, we've found that in fact many people have great gaps in their knowledge about sexual anatomy and functioning. We get hundreds of questions having to do with how long a penis should be and how to make it longer. Men and women wonder where the clitoris is, how erections work, when a woman is most fertile. There's a great concern about sexual normalcy: "Is my semen the right color?" "Is it normal to sometimes not want to have sex?" "Is it okay to be attracted to my aunt?"

Despite the assumption implicit in erotica, many people know little about the realities of sexual response. For example, unlike men, only a minority of women achieve orgasm through coitus alone. However, most erotica is written by men and is based on men's fantasies of women's sexuality. Thus we get a great many questions from men and women who think they're inadequate because the woman doesn't have an orgasm during intercourse.

We see profound problems arising from the failure of erotica to consider human emotional needs. Many men are looking for love, and don't understand why the mechanical approaches they practice so well aren't successful. Men often ask us for formulas ("How do I get a girl? What do I say? How do I know when one wants to go to bed? What are the foolproof ways?").

We detect concern over an absence of social skills. Many of our readers tell us they're shy and lonely. They often don't know how to meet people or make relationships work. Men especially seem to feel inadequate socially. They find it hard to make the first move, and constantly ask us what women want and how to tell if a woman likes them. Part of the problem stems from the mobility of this population. Educated professional young men find themselves working in cities far from families and childhood friends. Many work in largely male environments and are frequently transferred. Their careers call for social skills many find they lack.

Many sexual problems arise from sexual intimacy coming before emotional intimacy. Because partners don't know each other very well, they're uncomfortable expressing their preferences. Unready to communicate what they like, they commonly receive what they don't like, and fail to find physical gratification.

Often they feel pressured to enter into sexual relationships before they're emotionally ready, a source of sexual dysfunction. Premature ejaculation is a common problem. Many men ask about episodes of impotence. Women complain of failure to achieve orgasm.

The cruelest hoax erotica plays on our readers is the absolute omission of health considerations from the pursuit of recreational sex. The threat of sexually transmissible diseases comes as a shock, and our readers are preoccupied with reconciling the ideal pleasures of the body with the very real sores, discharges, and pain it might suffer.

There are innumerable questions we're asked about venereal disease, vaginal infection, prostatitis. There's concern particularly about herpes, and near-panic about AIDS. People nurtured on the fantasy of free and easy sex are having to come up against its unpleasant consequences. As a result, we get the feeling that after a number of years of casual sex there's been a retrenchment toward more selective sexuality.

Another common concern of readers is birth control. There's a feeling of frustration and anger that medical science has not yet come up with a totally safe, reliable, and easy-to-use method of contraception. In the meantime, people want help in understanding and using the methods at hand.

You'll find HSX's gay readership represented in the questions that follow. Gays who live in small towns are particularly isolated, and they seek through the computer the medical and social information that gays in larger communities find more accessible.

Even among heterosexuals, we've concluded that there may be a lot in the closet next door. Sexual activity in real life is often a far cry from supposed norms. There is widespread practice of behavior that clinicians may be too quick to declare pathological. For example, from our contacts on-line, we believe that many well-functioning heterosexual men practice some degree of cross-dressing. They feel no reason to take their concerns to therapists, and so transvestism may be more normative than many clinicians realize.

So where are Americans in the eighties? They're demanding a lot, absorbing a lot, willing to work a lot—but they've yet to reach the sexual ease and fulfillment that they think was promised to them.

The computer, we predict, will make a big change in that sexual picture. It offers a personalized approach to sex that can reassure people of their normalcy, educate them, and offer them support. It's private, and it can respond to secret concerns. It's current—it doesn't have to wait to "break into print" with the latest studies and facts. It can be much more direct and intimate than radio or TV.

We think the computer, as exemplified by HSX can be a useful guide through the sexual revolution.

What You'll Find in This Book

This book offers a sampling of material drawn from HSX. Scores of representative questions asked by subscribers are answered, many in consultation with a staff of experts who provide informed and sensitive medical and psychological advice. The questions and answers are the main portion of the book. The Index can help you locate particular subjects.

You'll also find stories about on-line communication based on actual incidents that have occurred on the service. These stories show how relationships have evolved, how they've changed lives, what they've meant to the people involved. Here you'll learn in detail about the profound impact this new medium can have on people's lives and how it can change both their perception of themselves and their understanding of what constitutes an intimate relationship.

Electronic conversations are transcripts of exchanges that have taken place on line. In these, people explore such issues as the meaning of computer communication in their lives, how computer friends compare to ''real'' friends, and the etiquette of on-line romance.

Confessional boxes are private thoughts and experiences readers have shared—from-the-heart statements people feel compelled to ''confess'' to the computer. Many readers preface these statements by saying ''I've never told anyone this before'' or ''No one in the world knows this, but . . .'' Sharing these thoughts anonymously and safely, many readers tell us, often helps them come to terms with aspects of themselves that have troubled or shamed them for years. In HSX, many of these confessions are displayed in a ''Letters'' section, prompting other readers to write, ''I thought I was the only one who felt that way,'' or ''It's nice to know I'm not alone.''

In all cases, identifying details have been changed to protect confidentiality.

Finally, we offer special thanks to our editor, Jean Arbeiter, for her sense, skill, and patience.

THE ELECTRONIC CONFESSIONAL

ON-LINE
RELATIONSHIPS

"What computer relationships are all about"

Barry: On the computer, it no longer matters what a person looks like, or what that person is physically capable of doing. People are represented by "pure thought." And in abandoning their physical component, becoming "letters on a screen," people are set free in a more profound way as well. Here one can dare to do anything.

Men who have existed for their entire lives as "Clark Kent" suddenly find that there really is a "Superman" inside them. They can attempt things in their imaginations—and in their interactions with other people—that would be out of the question for them in "real life."

I have been surprised by the kinds of double lives people live. Often, they're afraid of their real sexual feelings and can express them only on the machine. They think that if they released such feelings in "real life," others would shun them as perverts, or worse. The questions that they ask about sex reflect constant worry about their normalcy. And, without the computer, those questions would never have been asked.

I'm also constantly surprised to find that people are coming to HSX for companionship and computer "affairs" even though they have long been monogamous in real life. Perhaps it's true that people weren't meant to be monogamous, so when they free themselves from the ties of physical reality they can do what they were meant to do: explore sexual expression in all its many wonderful variations.

Here on the computer, all fantasy is safe. There is no physical contact, no fear of discovery or disease, no real risk. On the computer, one can make of the sexual interaction whatever one chooses. It can be as "real" as one dares, or simply pure fantasy—and thus no threat to one's sense of morality.

Computer relationships can be far more intense than one could believe possible. People become connected telepathically, and the relative anonymity of the contact makes it easy to pour out one's soul to an empathetic person on the other end of the computer network. One can speak freely, and without fear make revelations that would be unthinkable in real life. After all, one can persuade oneself that the other person isn't really there.

It has also been a surprise to me how easy it is to love people one has never met. Computer-connected love affairs also seem to be far more intense than those in real life. They start more quickly, freed of the physical bounds, and reach almost painful intensity in very little time. But this kind of relationship is hard to maintain. Either the people involved meet physically and things proceed or end from there or, realizing that there is no chance of any real future, the participants begin the painful process of withdrawing from each other.

On the computer, it's possible to meet people one would never come in contact with in real life: People from vastly different backgrounds, people from other countries, and people whose values and beliefs are entirely different from one's own. Again, the lack of physical contact provides security, and the "telepathic" communication provides instant access to new worlds. The ignorant and the learned communicate. The fundamentalist meets the homosexual without fear.

I think that as the world of computer communications expands, we will all come closer to being "one world" in a way that was never before possible. As long as we can read what we type to each other, we can communicate our real feelings on almost any subject, freed of the fears and prejudices that come from meeting people who are "different" from us in the flesh. Barriers must come down between people—and the computer is a good place to start.

———
30

To her co-workers, Julie Evans was the perfect example of a successful woman of the eighties. Though just 31, she was vice-president of her division in a very large corporation. She had gotten there by working twelve-hour days and plenty of weekends. Of course, she had the requisite MBA from a prestigious business school.

Yes, Julie had everything going for her. In addition to being bright and capable, she was always fashionably dressed and immaculately groomed.

There seemed to be no steady man in Julie's life, but she was so attractive that other executives were always trying to "fix her up" with friends. Sometimes they succeeded. But the relationships never lasted beyond a few dates. Julie claimed that her job came first. "I'm ambitious, I admit it," she would say. "It wouldn't be fair to the guy."

The night before Julie's thirty-second birthday, she slammed the door on Rob, a handsome and amusing attorney. He'd been pressing her to go to bed. When he grabbed her arm and tried to steer her to the bedroom, Julie screamed as if she were in a Victorian novel. "I've got your number, lady," Rob blurted angrily. "You pretend to have it all together, but you're just an old-fashioned iron maiden underneath."

After Rob stalked out, Julie looked at herself closely in the mirror. She could see herself looking at that mirror years from now— alone—as she rose higher and higher on the corporate ladder. Although Rob's words had been hurtful, they were on the mark. Julie was afraid of real physical contact with a man, and she was still a virgin. She was afraid to reveal her physical needs to anyone. Consequently, true intimacy was barred to her.

Not that Julie had no sexual desires. Quite the contrary. Many nights she masturbated while fantasizing that she was making love to a man. Masturbation had made her thoroughly familiar with her body's sexual responses. But she was terrified of letting a man know what she wanted. She was locked in a prison, confined by her fear of communicating openly with a lover. "I'll be a virgin until the day I die," she thought, not knowing that she would "lose" her virginity that very night.

Julie was too overwrought to sleep, so she logged on to her computer to do some financial research. After she finished, she switched over to HSX. She enjoyed following the conferences run by the single adult group, but she had never joined in.

Tonight, however, proved to be different. The discussion centered on fear of communication, so it caught Julie's attention. Julie was particularly impressed with the comments from a man named Tony. "Many people are afraid of getting hurt," he said. "So they don't let others know what they like sexually. But on the computer, people can admit what they like. That's what's so great about this machine. After all, no one can see you. You can open yourself up without fear of getting hurt."

To her amazement, Julie found herself saying, "I'm ready to open up, but I don't know where to start."

"I think I can help," replied Tony. "Let's talk." Julie knew that he was asking her to give her computer the command, "/talk." That command would put Julie and Tony in a private conversation where no one else could hear them. It would be like being in an electronic bedroom. Usually, when Julie was with a man she feared losing control. But tonight, in the electronic bedroom, she knew that she need go only as far as she wanted to. "Okay," she said. "I'll talk to you."

At first, Julie and Tony chatted about the difficulty they sometimes felt in revealing themselves to another person. Julie's father had been a very rigid man, and he had taught her to repress her emotions. If she didn't, he had told her, others would reject her. Tony listened as Julie told him things about her family she had never shared with another man.

After a while, Tony said simply, "I want to make love to you. Will you let me?" "Yes," Julie responded.

With those words, she entered a special fantasy world. Julie followed Tony's instructions to remove her clothing. She sat naked at the computer and absorbed his words. "I can see that you're beautiful. There's no reason for you to hold yourself back."

"I want you to touch me," Julie said.

"Tell me where and how."

"I like having my nipples rubbed. Rub them slowly and gently."

"That's just what I'm doing, my lovely Julie."

Julie closed her eyes and let her imagination take over. As she rubbed her own nipples, she could almost feel Tony's strong hand on her breast. She realized that she was beginning to flush and to breathe more rapidly.

Tony's next words were: "You look wonderful when you're aroused. What do you want me to do next?"

"I want you to lick my clitoris," Julie commanded. "Don't be too gentle, though. I need a lot of pressure."

"How did you know that I love to do cunnilingus?" asked Tony.

"I guessed you were the type."

While Julie rubbed her own clitoris, it seemed to her that the pressure was coming from Tony's tongue. She was more aroused than she had ever been in her life, when another message came from Tony: "Tell me when you want me to enter you."

In her mind, Julie let the oral sex continue for a few minutes longer. Then she said, "Now."

"I'm inside," was Tony's terse reply. The scene in Julie's mind shifted. She saw herself in her own bed with Tony. She moved rhythmically in his arms as he thrust deeply. Although she was manipulating her own clitoris, she could feel Tony's excitement mingle with her own. She could sense his weight on her body and his mouth on hers. She felt totally opened up. Suddenly, she had an orgasm, in an intense series of spasms.

When she recovered, she glanced at the machine. "Not bad for the first time, was it, lover?" were the words she saw. "Good night, beautiful lady." Tony had logged off.

As she fell asleep, Julie reflected that she was no longer a virgin. She felt as if she had truly made love with a man. On the computer, it had been easy to tell him what she wanted, to let her guard down. She had felt perfectly safe the whole time. She didn't really know Tony. And she didn't have to know him if she didn't want to. She was in control of the situation.

Tony, however, did not turn out to be a one-night stand. Julie had enjoyed the experience so much that the couple continued to conduct on-line adventures in loving. One day, Julie had enough confidence to suggest that they "voice"—talk on the telephone. The phone conversations were followed up by an actual meeting during Tony's vacation, and real-life lovemaking in Julie's no-longer-lonely bed.

Having sex in real life was not threatening, Julie found, because she already had "experience" with Tony. He knew all about her needs, and she was not hesitant about suggesting new ideas.

Although Julie's affair with Tony did not develop into a long-term relationship, she has since had such relationships with three men. One of them is now her fiancé.

Julie no longer feels frightened of sharing her physical needs with a lover. But she says that if she hadn't met Tony on the computer, she'd still be the "iron maiden" Rob complained about. "I lost my virginity on the computer," she says, "and it was a wonderful experience."

 ## "I found my dream lover on the computer"

Felicity had known Mike since childhood. In the small southern town where they grew up, both were "stars": he the captain of the state college football team, she the cutest cheerleader. They were a popular couple—everyone knew that they would be married someday.

Someday turned out to be right after Felicity got her teaching degree. Mike went into the construction business, and she took a job teaching English in the local high school.

At first, they were happy, but gradually discontent began to creep into the relationship, at least on her part. The trouble was that Mike remained strong and silent. His devotion to Felicity was unquestionable, but he had very little to say to her. Felicity, on the other hand, became more and more interested in literary subjects. After all, she taught English, and she wanted to tell Mike about the books she had read and about her classroom experiences.

Mike was willing to listen, but he had very little to say. He never read any books, and he didn't want to discuss his construction business. Marriage to Mike, Felicity thought, was like living with a pleasant, blank wall.

The same silence extended into the bedroom. Mike was a tender lover who seemed to understand, without talking, what Felicity needed to come to orgasm. But he was completely nonverbal. He didn't tell Felicity that he loved her, he whispered no words of endearment. After making love, he would simply roll over and go to sleep.

As she listened to Mike's rhythmic breathing, Felicity began to invent a fantasy lover. That lover was as physically adept as Mike, but he was expressive and intimate. He knew exactly how Felicity felt, not only about lovemaking but about great authors and current events as well. He would share Felicity's mind, as well as her body.

Mike bought Felicity the computer as a birthday present because, as he said jokingly, "it will give you someone to talk to." Felicity was indeed fascinated by the wealth of material she found on the computer service. She used it to bolster her lectures for her English classes.

When she discovered HSX, she was enthralled. Here were people who talked openly about their relationships and their sexual needs. Felicity became an eager participant in HSX discussions.

It occurred to her that HSX was the place to meet a man whose interests matched her own. If she couldn't have intimate, intellectual conversation at home, perhaps she could enjoy it on the computer. Felicity posted a "personal" which read: "Wanted, man who enjoys Shakespeare, Dickens, Hemingway, and Fitzgerald as much as I do. Age and appearance not important."

That's how the small-town English teacher hooked up with Ted, a recently divorced advertising salesman who spent a lot of time on the road and had few friends with whom to share his bookish interests. His computer meeting with Felicity turned out to be a true meeting of minds. It even seemed fated. At the very time that Ted answered Felicity's personal, he had just discovered the works of Edith Wharton—and so had she.

Their on-line encounters centered on literature at first, but gradually their talk became more personal. Ted was, indeed, the companion that Felicity had fantasized about. On the computer with him, she was able to express her intellectual loneliness. Ted told her all about his divorce and the desperation he had felt when his wife left him for another man. "I thought that we were able to tell each other everything," he recalled, "but all along, she was keeping her deepest feelings from me."

"I want to share your feelings," Felicity replied.

Felicity found that on the computer she could easily enter Ted's mind. In fact, communicating on-line was like talking mind to mind instead of person to person.

Ted gave Felicity the verbal stimulation she needed. She would glance up at the clock and be amazed at the amount of time she had spent talking with him. There seemed to be nothing she couldn't tell him. She found herself talking of how much she longed to be in bed with a man who would caress her with words as much as with his hands.

"I'm that man," said Ted. "Right now, I'm stroking your hair. I love the smell of it. Now, I'm moving my hand down over your breasts. The sight of them gives me great pleasure. Let me give you pleasure, too."

Ted and Felicity pretended that their meetings took place in a fantasy room. It was a library, with a glowing fireplace and a large, soft sheepskin rug. Sometimes they undressed one another on that rug and described what they were doing to each other's bodies.

"I'm getting ready to give you oral sex," Felicity would say. "I'm gently rubbing your penis with my fingers. Now, I'm putting it in my mouth. Can you feel the gentle nipping of my teeth? Those are love bites I'm giving you. I'm moving my mouth up and down, now I'm moving more rapidly. I can feel your penis becoming larger, larger, larger. It's filling up my mouth. I feel so excited."

Mental sex was only one activity that Felicity and Ted pursued. Their "library" contained all of the books that they had read and many that they intended to read. Near the fireplace, there was a large pile of magazines. They even had a make-believe VCR in which they showed films they both enjoyed. Some nights, they simply discussed books or movies, not bothering with "sex" at all.

The computer relationship was the most intense relationship that Felicity had ever known. All of her daily experiences were enhanced because she knew that she would be sharing them with Ted. It felt incredible to be mentally linked to another human being, especially one who responded totally to her inner world.

Felicity and Ted began to talk about meeting. A real-life love affair, they thought, was inevitable because they were soul mates. But Felicity feared hurting Mike, who, after all, had been totally faithful to her. As time went on, though, Ted seemed to take priority in her thoughts. He occupied almost all of her waking life. Without Ted, she could not be whole. They had to meet.

One night, Ted told Felicity that he was scheduled to spend a week in Atlanta. Would she meet him there? Felicity agreed. She told Mike that she would be visiting an elderly aunt in southern Georgia. Then, she and Ted arranged their schedules so that their planes would land in Atlanta at approximately the same time.

Throughout their "love affair," Felicity and Ted had never described their physical appearances. There was no need to; they believed that they knew each other intimately without such details.

Now that they were to meet physically, they were sure they would recognize each other by intuition. "I'll know you by the intensity I'll feel when I see you," Felicity said. Ted agreed.

But the meeting did not prove to be as simple as they had expected. When Felicity arrived in the airport lounge, she found hundreds of people milling about. She was shoved one way and then the other. She tried to stand perfectly still and look at the faces of the men around her, but Ted was not among them, she was sure. Nowhere did she see that glow of intelligence, or the recognition she expected. Where was the inner spirit she had come to know on the computer?

Ted had the same problem. Could any of these women really be his Felicity? He felt no electricity in the room.

Finally, after two complete planeloads of passengers had left the waiting room, they realized that they were the only two people who had been in the room for over an hour. They approached each other hesitantly. Nothing about the other person seemed familiar. Disappointment, rather than rapture, was the first emotion they experienced upon meeting.

Both of them were physically attractive, but that didn't seem to matter. Missing was the mental magic they had expected to find. In person, Ted's conversation sounded banal. Felicity's laughter was forced instead of natural. They groped for words so much that they might have been two people on a blind date.

Perhaps they could regain the relationship if they sat down and talked. Over drinks in the airport bar, however, their conversation only became more and more stilted. They didn't seem to be able to relax as they could on the computer.

Felicity and Ted cancelled their plans to go to a hotel. It would be pointless, they thought, for they felt no sense of intimacy. "Maybe we can find each other again on the computer," Ted suggested sadly as they parted. Felicity wept throughout the flight home.

As Ted had proposed, they did get together again on-line, but the spell had been broken. After a while, they stopped their computer meetings altogether.

Felicity was distraught for a long time afterward. "I have nothing to live for," she thought. Even today, she feels that part of her inner spirit is dead.

Will Felicity ever find a man who can share her soul? She's not sure, but she has decided to seek him in the real world instead of on the computer. She and Mike have filed for divorce, and Felicity is studying for her master's degree. On a university campus, she thinks, she may meet a more suitable mate. But she still cherishes that dream library where she and Ted—in absolute safety—made love to each other's minds.

 *"I have to wear women's clothing"*

Terry was eight years old when he realized that he could become sexually aroused by dressing up in his mother's clothes. He loved the soft texture of her sweaters and the smooth feel of her silk blouses next to his skin.

At first, Terry's mother thought that she simply had a son who loved to "dress up." But then, she became frightened. She forbade Terry to touch her things and explained that boys' clothing was more appropriate for him.

The forbidden aspect of the dressing-up made it more exciting. Terry waited for his mother to go out so that he could sneak into her room and put on her garments. When he saw her car pulling into the driveway, he'd hurriedly undress—his heart pounding— and pull on his own clothes.

By the time he was in high school, Terry knew that an important part of his sexual identity lay in wearing women's clothes. He wasn't a homosexual, he was sure. He was strongly attracted to women and dated a lot. He found, though, that he was even more sexually aroused by women when he was wearing feminine clothing.

Secretly, Terry purchased lingerie for himself, telling the saleswomen that it was for his girlfriend. Under his rumpled blue jeans, he frequently wore women's panties. When he was home alone, he dressed up in a bra, panties, panty hose, a dress, makeup, and a wig. Just putting on the outfit aroused him so much that he often masturbated. As soon as he heard his family returning, he buried his beloved clothing beneath a pile of T-shirts in his drawer.

Terry was a transvestite, a man for whom wearing women's clothing is a turn-on. Like most transvestites, Terry kept his preference to himself. He was both excited and frightened by it.

After Terry got married, he continued to keep his secret. He longed to wear women's clothing while making love to his wife, but she was straight-laced and he was in terror of losing her. Many transvestites do confide in their wives and some women accept the cross-dressing. Some even share their clothing with their husbands. Terry, however, was afraid to reveal his secret life to anyone.

Instead, he accumulated more and more women's clothing and kept it locked in a trunk in the basement. He followed the fashion magazines avidly. He'd buy an entire wardrobe of outfits and then, in disgust, toss them all out, resolving never to wear women's clothing again. It was impossible for him to keep that resolve. The more anxious he was to give up cross-dressing, the more it increased its hold on him. Sooner or later, Terry would yield to his desire and purchase a new collection of clothes.

He lived in constant fear. Suppose his wife, his children, or his business associates found out? Sometimes, though, it made him laugh to think what his co-workers would say if they knew that at least one member of their management team was wearing panty hose under his pin-striped suit.

Usually, Terry could only dress up completely when he was alone in the house. But when he traveled on business, he took his female clothes with him. In a strange hotel room, in a strange city, he felt more secure about slipping on one of his favorite ensembles, making up his face, and topping the outfit with a chic wig. Then he'd venture out on the streets, walk around for a bit, and go to dinner or the theater. He was exhilarated by these outings, and felt he could set his real self free at last.

Sometimes the evenings ended with Terry in the arms of a prostitute. Prostitutes, he had found, were accepting of his practice; indeed, they were happy to go along with the game. The problem was that being with them left him feeling more lonely than ever.

One night, when he was in a small Western city, disaster struck. A policeman spotted Terry strolling through the downtown shopping area and became suspicious. Terry seemed awfully tall to be a woman, and his shoulders were broad and masculine.

"I want to talk to you, fella," the policeman barked. Terry's heart almost stopped beating. Within seconds the cop had him pinned against the wall, while an interested crowd looked on. The policeman explained, not too unkindly, that the state had a law against cross-dressing. By now, the tears were streaming down Terry's face and his silken underpants were damp. He could see his life crumbling around him.

"Listen," the policeman said softly, "why don't you just go home and get dressed right—you know what I mean? I don't want to see you around here again."

Terry moved so quickly that he didn't even remember getting back to the hotel room. He double-locked the door and pushed the bureau up against it. He felt totally exposed. The worst part was that there was no one with whom he could share his terrible feelings.

Several months later, Terry found a place where his secret self could live: the computer. While exploring HSX, he discovered that there was a special support group for those with alternative sexual life-styles. Here, transvestites could openly discuss their fantasies and experiences.

After Terry became a member of the group, he realized that there were many men like himself. Today, he feels that his needs, while not usual, are harmless enough. Why shouldn't he be able to express them and share them with others? On the computer, he can do just that—without fear of discovery or arrest.

He has also found that the more he discusses his desire to wear women's clothing, the less compulsive he feels about doing so. He can enjoy his fantasies without being totally enslaved by them.

It's usual for members of the support group to use women's names to converse with one another. Terry calls himself Teresa and he uses the initials "tv" after his name, as the others do, to indicate his sexual interest, transvestism.

On the computer, tvs are able to talk about their passion for women's clothing in minute detail. They share tips for shopping and for shaving one's legs and underarms with ease. Many, for the first time, have found willing ears for their concerns.

Here's an example of the kind of conversation Terry enjoys with other tvs. The topic for the day is "My favorite outfit."

Liz: I just love anything made of angora. There is nothing that makes me feel more feminine than the softness and fuzziness of angora. My favorite outfit consists of a rose-pink angora sweater and a white angora skirt. The sweater has long sleeves, ribbing at the hem and cuffs, a jewel neckline and a pearl button closure at the back of the neck. The skirt is rather straight, though not tight, and comes down to about mid-calf. Under this outfit, I usually wear a pink, underwired, lacy Vanity Fair bra in size 38B, matching high-leg panties with lace, a matching pink half-slip with side slits and an inch of lace along the hem, and beige panty hose. Then I wear white sandals, with ankle straps and three-inch heels.

For jewelry, I usually wear a single string of pearls and filigreed gold earrings. Needless to say, I always wear full makeup with this outfit: foundation, powder, rose-colored blusher, pink eyeshadow in three shades, mascara, and red lipstick. My hair is usually blonde, coming to my shoulders. It's wavy with bangs.

Now I'm walking down the runway and looking at you. How do I look?

Teresa: You look pretty terrific. I'd just shorten the skirt somewhat. I myself prefer my skirts to be knee-length. For a new look, why don't you try some patterned or glitter-type panty hose? Otherwise, don't change a thing.

Liz: Thanks loads. Actually, I do have some patterned hose. I've got a beautiful pair of white lace that cost me eighteen dollars!

Eva: I'm dressed up today because my mother is out. If the cat's away, the mouse will play. So, although I won't stay dressed up much longer, it's fun while it lasts.

Sorry, I didn't have time for full makeup. I had trouble shaving so I just have on red lipstick. I also put on my fun outfit. I started with Sheer Energy panty hose, then topped them with darkish red opaque tights. For underwear, I'm wearing a body suit stuffed with bra forms and an extra bra to shape me better. The main part of the outfit is a black jump suit. It has very short sleeves and a drawstring waist. Finishing off the outfit are a wig, black flats, and white button-style earrings. That's my play outfit for today.

Teresa: Sounds like an outfit to really have fun in. I hope you are. Let's talk more later.

Graham had no doubts that he loved his wife. He'd been married to Madeline for over thirty years, and he'd never seriously looked at anyone else.

The catch was with the word "seriously." Although he had never actually contemplated going to bed with another woman, lately he had found himself looking and fantasizing a lot more than he ever had before.

Sex in Graham's marriage had become routine, and Madeline resisted all of his suggestions for spicing it up. Sometimes she would try a new position, or a new place to make love, but she always wore a somewhat martyred expression, as if she "owed" the dirty deed to Graham but wasn't enthusiastic about it herself.

The truth was that Madeline's sex drive and sexual interest simply didn't match Graham's. There were all kinds of sexual experiences that interested him. He wanted oral sex much more often than Madeline was agreeable to it, and he even wondered what it would be like to engage in group sex. Madeline would never go for that, he was sure.

Still, as much as Graham's mind wandered, he didn't want to jeopardize his relationship with Madeline. She trusted him. They had loved and supported each other through many hard times. Surely all of that was more important than the more active sex life he envisioned.

Fran, a co-worker, introduced Graham to HSX when she had the couple to dinner. Computers were relatively new to Graham and Madeline, so Fran showed them how to log on. Then she showed them HSX and the on-line conversations that were taking place. Fran told them that she found HSX exhilarating. "At any hour of the day or night, I know that I can find an interesting conversation. Sometimes I have problems with my relationships, and I want to know how other people are handling the same kinds of things. Also, you can meet very interesting people."

That last sentence stayed with Graham. Maybe, on the computer, he could meet some of the "very interesting people" he had fantasized about.

Graham purchased a computer and became an HSX enthusiast.

He began to realize that the computer could fulfill his fantasy of having an extramarital relationship. At the same time, it could prevent him from being unfaithful to his wife.

On-line, Graham took part in discussions in HSX's support group for couples. He was able to develop a relationship with Annie, a woman who was very close to divorcing her husband. She wasn't quite ready to make the move yet, and she needed a special kind of friend, someone who would listen to problems, yet make very few demands on her. In return, she fulfilled Graham's fantasy of having a mistress.

So now, each night at 10:30, Graham kisses Madeline good night and switches off the light. When he's sure that she's asleep, he slips out of bed and goes downstairs to the den, where he switches on the computer.

There, Annie is usually waiting for him. Both of them give the computer the command "/talk" to obtain privacy.

"You look gorgeous tonight, lover," Graham might say. "What are you wearing?"

"A black lace crotchless teddy and Opium perfume. I can't wait to make love to you."

"Well, I'm not wearing a thing. So do whatever you want."

Sometimes Annie describes the way she's massaging Graham. Sometimes Graham tells Annie that he's preparing to enter her from the rear. They vie with each other in describing the sweet-tasting substances they're rubbing on each other's genitals. Whipped cream is a favorite of Graham's, and sometimes it seems to him that he feels it on his penis, even though Annie is a thousand miles away from him.

Annie and Graham can imagine that they're making love in the middle of a crowded railroad station, or they may be in a quiet glen. Sometimes they bring in a crowd of other people they've invented and create an on-line party. Whatever they do, they feel that it's absolutely harmless. Being on the computer is fantasy-time for both of them.

Their relationship gives Graham the feeling that he's expanded his sexual repertoire. It gives Annie some respite from the problems of her marriage. Both are satisfied without forming a real-life extramarital relationship. In fact, such a relationship would defeat their purposes. It would cause more emotional chaos for Annie, and it would fill Graham with intense guilt.

Of course, both Graham and Annie know that their alliance can't last much longer. Annie will soon have to come to some real-life decisions about ending her marriage. Then, she'll want to start rebuilding her life. In all probability, Graham won't fit into her plans, though the pair may be able to maintain an on-line friendship of some sort.

When Annie leaves him, Graham will have to decide whether he wants to find another computer mistress. Perhaps having fulfilled the fantasy once will be sufficient. Graham isn't planning that far ahead. "All I know is that I'm happy with things the way they are," he says. "The computer has allowed me to commit adultery without being unfaithful to the woman I love."

"I want to be safe from disease"

Jeff went to college in the early seventies at the height of the sexual revolution. Hang-ups and repression? They were for dodos. Everyone knew that sex was good for you. If you didn't develop your sexual dimension, you were cheating yourself.

It didn't matter how many people you had sex with. Advances in birth control had made intimate relationships entirely safe. There seemed to be no reason to be reticent about going to bed with a fairly new acquaintance. By the time he graduated, Jeff figured he'd had sex with at least half of the girls in his dormitory. In fact, he had come to the point where he expected a date to end in intercourse.

After college, Jeff continued his freewheeling life-style. He was an engineer, and as an upward-bound professional he had no trouble attracting women. Jeff was a familiar figure at the neighborhood singles' bars. Rarely did he leave one without a young woman on his arm. Some mornings he woke up to find himself in bed with two women. Occasionally, he indulged in group sex. He was, as he told both old and new acquaintances, "open to experience."

But a few years ago, Jeff's sexual life became less exciting. For one thing, he was experiencing an emotional deadness. He felt no sense of commitment or closeness to anyone, in spite of his sexual expertise.

Then there was the talk of herpes that he seemed to hear all around him. The talk didn't really get to him, though, until Jeff barely escaped a herpes encounter of his own.

At an engineering conference, he met Helene and invited her to the hotel bar. Although Jeff sensed that Helene was just as strongly attracted to him as he was to her, she seemed to be trying to keep aloof from him. She smiled pleasantly enough, but she wasn't engaging in that casual body contact that indicates, "You turn me on."

Still, Jeff thought, he could be reading her signals wrong. So he invited Helene to his room. She agreed, but when they got to his room, Helene began to behave peculiarly. She undressed a bit too hurriedly, for one thing. Then she seemed more and more distant, as if she wanted Jeff to forget the whole thing.

When Jeff touched her, Helene burst into tears. "I can't go ahead with it," she said. "You're the most terrific guy I've met in a long time. So I've got to level with you. I have herpes. I'm asymptomatic now, but even so, it's not safe for us to make love. I couldn't take the chance of infecting you."

After that experience, Jeff's concern about his sexual encounters turned into full-blown fear. It wasn't the devil who was trailing him, it was venereal disease. He read that VD was epidemic in the United States, second only to the common cold in frequency among contagious diseases. Gonorrhea was on the increase. There was also a host of lesser-known dangers—herpes, chlamydia, and trichomoniasis, to name a few.

Every day, it seemed to Jeff, he heard about some friend who had one disease or another. He began to feel betrayed by the sexual revolution. The Garden of Eden, it seemed, was full of poisonous plants. Jeff was angry. This was not what he had bargained for when he was taught that sexual experience was the pathway to fulfillment and freedom. The ground rules had totally changed.

Jeff found himself looking at the women he met with suspicion. Did this one or that one have herpes? How could he possibly find out? He knew that he couldn't count on all of his partners to be as honest as Helene.

One night, Jeff discussed the situation with Bill, an attorney friend. "I'm going to ask the next girl I meet to sign a contract stating that she doesn't have herpes before I go to bed with her. Can I take her to court if she lies to me?" At first Bill laughed. Then he realized that Jeff was serious. "Your contract might stand up," he commented, "but what good would it do if you contracted herpes? If you're so concerned, you should stop going to bed with women altogether."

Jeff took Bill's facetious words to heart. He decided not to make love to anyone until he was ready for a long-term relationship. The decision amazed his friends: Here was a man who had scarcely been alone a night since his adolescence.

But Jeff finds celibacy more comfortable than living with fear. "I'd become absolutely paranoid about approaching a woman I didn't know," he says. "All my old mechanisms for meeting people didn't seem safe anymore. I felt that it was just a matter of time before I, too, contracted VD. There was even the possibility of AIDS if one of my singles' bar pick-ups had been to bed with a bisexual male."

Jeff's solution has been to join HSX. In "The Singles Club" support group, he feels that he can enjoy relationships without the danger of disease. "I've gotten to know several women well through the computer, even though I haven't met them in person. When I'm on-line with them, especially privately, I can tell them my deepest thoughts. I can even come on to them if I want to, just as I would in a bar. The only difference is that it doesn't end up in bed. But would you believe it—the peace of mind is worth it to me.

"As to the future, I'll have to see how things go. Probably, I'm going to want real-life sex sooner or later. I may even want to meet some of the women I've met on-line. After all, the computer has allowed me to get to know them quite well. It provides time for the preliminaries that simply aren't possible in a bar or on a blind date.

"But I'm just not ready yet for a serious relationship. I think I'm waiting for the sexual revolution to get its act together, disease-wise. Maybe that's unrealistic. But for the present, I prefer my relationships to be on-line. In my imagination, they can be as sexy as I please, and I know that I won't wind up in the doctor's office."

"I've never told this to anyone before"

Some people carry secrets they think they can divulge to no one. They go through their daily lives bogged down by guilt and unhappiness. For such people, the computer offers a place to lay down the burden—if only briefly.

One day, a discussion was taking place among the members of HSX's "For Women Only" support group. The subject had turned to combining a career with motherhood. Angela and Leslie were involved in an intense interchange with Beth, the mother of a six-month-old, who had recently accepted a job.

"I've advertised for a baby-sitter," Beth said, "but there seems to be something wrong with everyone who shows up. My husband says I'm being too picky. All of the applicants make me nervous, though."

"You're probably feeling anxious about leaving the baby," Angela pointed out. "Why don't you ask one of the potential sitters to spend a day with you and the baby so you can see how she takes care of him. You can learn a lot about a person in one day."

"A friend of mine takes her child to a woman who baby-sits in her home," Leslie chimed in. "There are a couple of kids there the same age as my friend's and it's worked out real well."

After a while, Leslie confided that she partly envied Beth's decision to go back to work. "Some days I get so bored at home that I could scream. I miss all the people I used to talk to in the office. And, of course, there's the money. But I do want to be here when the kids get home from school. I love them, even though I feel like killing them sometimes."

Suddenly, a new voice intruded into the conversation. "I hate my baby all the time," it said. "I'm sorry that I had her. I hate being tied down night and day. Everyone says how lucky I am to have such a beautiful child. I don't feel lucky at all. I smile and pretend to be the perfect little mother. I tell my husband how happy I am. Actually, I wish I was dead. No one knows how I feel. I've never told this to anyone before."

The log-on chart showed that the newcomer's name was Frances. After she had spoken, Frances disappeared as quickly as she had come. Beth, Angela, and Leslie tried frantically to reach her.

"Frances, I'd like to talk to you more about this. You must feel terribly lonely. I used to have feelings like yours right after my son was born. I think I can help you deal with them. Please, please show us that you're listening," said Beth.

But there was no further word from Frances. She had unburdened

47

herself—then, frightened by the vehemence of her confession, had backed off.

Was she still listening, as she had apparently been doing all along? There was no way to tell. On the computer, any group discussion can be overheard by many silent listeners, or by none at all.

The other women wanted to help Frances. But perhaps she didn't want to explore her feelings any further. Perhaps she had just wanted to share them and then return to her desolate silence.

"On the computer, I'm just like everybody else"

Born with a severe birth defect, Stephen has been in a wheelchair for almost all of his 28 years. He has no movement in his legs, and his arm motion is limited, but he is capable of using his hands. In spite of his handicap, Stephen was always an outstanding student. With help in getting around, he could adjust to life in an ordinary school.

Although Stephen was used to depending on other people for help, he found that he could rarely count on them for friendship. Very few of his classmates ever tried to become close to Stephen, unless they themselves were handicapped.

At puberty, Stephen developed sexual urges that were as strong as anyone else's. He dealt with them, as many boys his age did, through frequent masturbation, but he longed for a real relationship with a girl.

In spite of his handicap, Stephen tried reaching out to girls. Most girls, he found, would date him once—sometimes twice—because they didn't want to hurt his feelings. But the relationships always remained in the category of friendships. Few girls wanted to kiss Stephen, much less make love to him as he wished that they would.

Stephen's longest-lasting high-school relationship was with Karen, who had a strong maternal instinct. Her concern for Stephen was clearly based on sympathy, even though she was eager to perform fellatio on him. Still, as much as Stephen enjoyed Karen's attention, he always had the feeling that she looked upon him as an unfortunate child, rather than a lover. Karen was better, though, than the girls who dated Stephen as a curiosity. Their uneasiness was always apparent.

Eventually, Stephen gave up the idea of genuine relationships. People saw only his handicap, he thought, not his real self. In college and graduate school, Stephen majored in finance and devoted most of his time to his work.

After graduation he was hired by a securities firm, and once again his life was dominated by work. His sex life was relegated to fantasies. But in these fantasies, Stephen was free of his wheelchair. In them, he was strong and athletic. His sexual prowess was such that he had no trouble attracting women. In fact, he had so many that he scarcely knew which one to make love to at night . . .

In reality, Stephen could find no way to assuage his loneliness. He lived at home with his parents, and as they grew older, Stephen realized that he would soon be totally alone.

Stephen had a computer in his office. He frequently logged on to CompuServe to check out corporate reports. One evening, when he was working late, Stephen explored the other parts of the service and came upon HSX.

He found himself engaged in conversation with a young woman who, like himself, was in her twenties and eager to meet guys. Stephen truthfully told her that he was a securities analyst, and she was impressed.

"Looks?" she said suddenly. Stephen was caught by surprise. He hadn't realized that she didn't know what he looked like. This was the first time in his social life that he had encountered another human being who didn't know he was handicapped.

Stephen scarcely hesitated. "I'm five feet, eight inches, blonde hair, brown eyes, 162 pounds, and interested in athletics." All of those facts were true. Stephen had simply left out the fact that he was in a wheelchair.

Within a few days, Stephen purchased a computer of his own and became a CompuServe subscriber. He realized that the computer gave him an amazing new option. It enabled him to live out the fantasy that he was not handicapped. He had learned how much being handicapped could impede the formation of relationships. On the computer, there was no wheelchair.

On the computer, Stephen could walk; indeed, he could run. He introduced himself to people as a tennis player. The women he talked with quickly learned that Stephen played a strong game of tennis and was currently thinking of taking up golf. He was a hiker, and a member of the Sierra Club.

Of course, Stephen made no attempts to meet these women; they were friends rather than potential lovers. Nonetheless, expressing the fantasies gave Stephen a tremendous sense of freedom. At last, his handicap was unimportant.

As Stephen achieved some degree of acceptance from others, he began to feel less anxious to cover up his handicap. He could, he found, be Stephen with or without a handicap—as he chose. It was the freedom of choice that was liberating. Surprisingly, it made Stephen more accepting of his handicapped self.

Today when Stephen meets people on the computer, he plays the conversation by ear. He may tell them that he's handicapped, and he may not. "I need to be open and honest, yet I will not blab my handicap out in search of sympathy. I tell my computer friends about my handicap when and if it's pertinent to the conversation. HSX gives me the option of discussing my handicap when and if I want to. I no longer have to pretend that I'm something I'm not."

The computer has given Stephen an increased sense of self-worth. Now he finds that he can reach out to other people in the real world, too. He has joined a support group for young adults with handicaps, and he has increased the number of his non-handicapped friends, too.

 ## *Desperately seeking Judy*

The Westchester County, New York, house that Judy and Hy lived in seemed to go on forever. There were six bedrooms and two family rooms. There were also two acres of land—kept beautifully manicured by a team of gardeners—and a glistening swimming pool.

Still, Judy was not satisfied with her life. Even with three beautiful children, ages nine, eight, and four, she was dissatisfied. It seemed to her that she was going out of her mind with boredom. Often, she felt like the housewife in the film *Desperately Seeking Susan.* "But I don't have the guts to move to New York City and look for my real self," she thought.

It was ironic because Judy's current material success was the only thing she had dreamed of as a child. She came from a dismal neighborhood. Watching her parents struggle with poverty, Judy had determined that she would do better in life.

The solution to her economic problems came more quickly than she had hoped in the person of Hy, a manufacturer for whom Judy's father worked briefly. One day, when Judy stopped by the factory to see her father, she encountered Hy. That was the start of a whirlwind courtship that ended with an elaborate wedding—paid for by Hy—and the move to the suburbs.

At first, Judy told herself that her dreams had come true. Her parents were secure. In fact, Hy bought a small apartment for them in the same suburb.

Within a few years, however, Judy had to admit that her relationship with Hy was troubled. He was a brusque individual who treated her as if she were simply another one of his possessions. He had acquired her as he acquired any other object that suited his fancy. Now he felt that he no longer had to pay much attention to her. Hadn't he provided her with enough toys to keep her happy?

It was at her son's tenth birthday party that Judy realized how unhappy she really was. The festivities were scheduled for the evening so that Hy could attend, but he never showed up. He didn't even bother to call with his usual excuse about having to close a business deal.

For some time, Judy had suspected that Hy was seeing other women. No one, she reasoned, could possibly have to close that many deals. His lovemaking had long since become perfunctory and more and more infrequent. Still, Judy hesitated to voice her suspicions to Hy. After all, she was dependent on him for her physical comforts. There were the children to think about—and her parents, too. Why couldn't she simply accept the arrangement as it was and stop longing for a more intimate relationship?

The party, however, put an end to Judy's struggle to accommodate herself to the status quo. When the guests had gone, she sat in the family room, too agitated to sleep. Then, in the corner of the room, she saw a large package. It must have been delivered in the afternoon and accepted by the maid. Inside, Judy discovered a top-of-the-line computer. It was an inappropriate gift for a child, but Hy always confused extravagance with devotion.

In the morning, Hy murmured apologies for missing the party. Judy accepted them in her usually submissive manner. Emotionally, though, she was beginning to disassociate herself from Hy.

During the months that followed, Judy mastered the use of the computer. She discovered various services—and HSX—and found that HSX gave her a chance to establish on-line relationships with other adults.

"For me, being on HSX was like escaping to another world," Judy recalls. "I felt that I was in a gigantic singles bar, full of potential lovers. On the computer, I wasn't a boring housewife with very limited life experiences. Instead, I could be a sophisticated woman of the world. Talking on the computer liberated me from my everyday personality."

Gradually, Judy's computer conversations began to center around one man: Vince. He was a 22-year-old noncommissioned naval officer stationed in Virginia. He had joined the navy, he confided, to escape a background of oppressive rural poverty. He had seen some of the world and was anxious to learn more about it from his computer friends.

Judy came on to Vince with a fantasy: She was, she told him, very much a free spirit. In fact, she was a 24-year-old artist (her real age was 32) who lived in New York City and frequented trendy bars and discos. Her art was commercially successful, and she made a good living at it. In fact, she was a recognized "character" in the art world—she was sure that Vince might have read about her in various magazines.

It was easy to fool Vince. He was so young and so eager to believe that he had chanced upon a dramatic figure, a woman who knew what life was all about. As the two chatted alone, Judy beguiled him with stories. She told him how she had lost her virginity at an artists' colony in the Berkshires with one of the most renowned exponents of the Pop Art movement. She described her most recent roommates, a pair of gay lovers who ran an off-Broadway theater company. When she had to be off-line for a week because she accompanied Hy on a business trip to London, she told Vince that she had been in the hospital, recovering from the effects of an orgy that ended in violence.

Vince was thoroughly hooked, but in time Judy began to become hooked herself—by Vince's open nature and sincerity. She felt horribly guilty, but she couldn't bring herself to end their relationship. Vince had come to mean a great deal to her. Besides, she reasoned, she was entertaining him and giving him exactly what he wanted. The computer was linking his vivid imagination to her own.

Eventually, Vince had had enough of stories. He wanted to meet Judy in person. He wanted to see her Greenwich Village loft and meet her friends. He wanted to make love to her in person. He told Judy that he knew her as well as he could on the computer. Now he had to experience her in real life.

He was scheduled for a leave, he told Judy. Where could they meet in New York? Judy knew that she should log off the computer for good. But she couldn't find the courage to do so. Vince was a major figure in her life. She longed to meet him, even if meeting him meant losing him. After all, perhaps they wouldn't like each other in person. It was worth finding out.

Judy told Vince to meet her at a popular art gallery in Manhattan. For the meeting, she purchased an outfit of rhinestone boots, tight pants, a flamboyant satin shirt with matching scarf, and a second-hand fur coat that trailed on the floor. She spiked her hair and applied fake eyelashes, but when she met Vince, with his wide-open eyes and regulation crew cut, she was ashamed of her disguise.

Over coffee, she admitted the truth. She was a person whose adventures in life were even more limited than his own. Also, she was ten years older than he was.

Vince simply laughed, as if she were telling him one more delightful story. He said that he had fallen in love with Judy, no matter who she was. "It's time that I really got to know you," he said, and he took her to a hotel where they made love, free of pretense.

After Vince returned to Virginia, the pair discovered that one day together was not enough. On the computer, they began to talk about being together permanently—and this time, Judy was not dealing with fantasy. She felt that she simply couldn't lose Vince. She looked upon him as a gift of fate—a gift from the computer. One night, she revealed the whole story to Hy and her astonished parents. She wanted a divorce, she said.

Judy's mother sided with Hy in his successful attempt to gain custody of the children. She was convinced that Judy had lost her mind. In fact, Judy's mother said that the computer should be named as corespondent in the divorce.

Today, Judy lives with Vince on the naval base in an apartment she would once have scorned. Due to the divorce settlement, she's not without resources, but Vince insists that they live on his salary—and Judy really doesn't mind. She's enrolled in a fine arts program at a nearby community college, and she feels that someday she'll have the artistic career that was once hers only in fantasy.

The painful part of her decision to marry Vince was the loss of her children. She sees them one weekend a month and talks with them on the telephone almost every night.

But there's nothing painful about her real-life relationship with Vince. He proved to be as truly loving as he seemed to be. In marrying him, Judy believes that she has found the "real" Judy at last.

"Only my computer friends know that I'm gay"

For gay adolescents, life can be bleak and frightening. They feel thoroughly isolated and unable to confide in anyone. It's rough enough to be a teenager, but the gay adolescent must also deal with the knowledge that he is "different." Instead of being turned on by the opposite sex, he's turned on by his own.

Many gay adolescents report that, even as children, they somehow "knew" they were gay. But in adolescence, when problems of sexual identity become paramount, they must confront that gayness. Whom can they tell? They're terrified of confiding in their parents, and they fear—often rightly—that straight friends would desert them if they knew the truth.

David was a 17-year-old who had always felt a strong attraction to his own sex. Of course, same-sex crushes are normal in adolescence. But David's entire identification was with other men. When he thought of a loving relationship, he thought of being involved with another boy.

David lived in a small town where social events focused on the church for adults and the high school for teenagers. On Friday nights, the entire town turned out for the high school basketball game. On Saturday nights, there was often a school dance.

David's friends thought of him as "shy." Sometimes he had to be coaxed into asking a girl to the dances. Usually, though, he did so, and after the dance, he even engaged in the requisite amount of parking. David dated so that he could keep up his relationships with his friends and talk about girls with them. Even as he kissed a girl in the backseat of a car, he fantasized that he was actually kissing his best friend.

David's parents, both of whom were active in the church, were glad that he was on the shy side and dated such sedate girls. David imagined that both of them would die instantly if they knew what was going through his mind.

One of the things that was frequently on his mind was an incident that had happened in the sixth grade. David had shared a desk with Raymond. One day, David observed that Raymond was quietly masturbating. He had placed a book on his lap and beneath it, he was manipulating his penis with his thumb and forefinger.

David found the sight tremendously arousing. Raymond pretended that he hadn't noticed David's interest and went on with his activity, staring straight ahead. Slowly, David moved his hand so that it was closer to Raymond's penis. Then, without saying a word, he put his hand on Raymond's penis and took over the rubbing for him.

Although Raymond and David took turns at masturbating each other for the rest of the school term, they never discussed it. At night, David dreamed of kissing Raymond and holding him in his arms. He would wake up in the morning to find his sheet damp with semen. After Raymond moved away, David longed to repeat the masturbatory experience, but he didn't dare mention his longings to his friends.

David met a youthful gas station attendant one summer at the beach. The young man took him out to a dune and made tender love to him. David found sex with a man just as fulfilling as he had always suspected it would be. He found, though, that the confirmation of his homosexuality made pretense at home even more difficult. He longed to ''come out,'' but he was terrified.

David found solace when he joined the Gay Youth support group on HSX. Here, he discovered, there were other young people who, like himself, were isolated and frightened. He had imagined that he was the only person in the world who was ''different'' from the majority. Now he learned that there was an entire group of teenagers with whom he could relax and be himself.

David discovered that for the kids in the group, ''coming out'' was a major problem. They frequently talked about whether they should do it, how to do it, and when.

Eric described the situation well: "I've met this great guy, **Todd.** I'm sixteen and he is, too. We've been seeing each other for **about** three months now. He's just great. I love him so much.

"The problem is that at times we have to lie to our parents, especially mine. It doesn't feel right. I really want to share what I'm feeling with my parents. So how should I handle it? Should I just walk in holding Todd's hand and say, 'This is it'? Should I write them a letter or what? I've never told anyone yet, so I'm sorry if this question sounds kinda dumb."

Usually, the other kids advised someone like Eric not to come out. Coming out during adolescence, when a teenager is financially and emotionally dependent on his parents, can cause huge problems. Frequently, parents do not accept the news well. They become frantic and attempt to discipline or "cure" the child. The revelation can create estrangement rather than closeness. Generally, parents can accept the news of a child's gayness much better when he is grown and out of the house.

Victor, another member of the group, told Eric: "I don't think you should tell them until you are at least eighteen because some parents can't handle that. I know mine wouldn't be able to. Parents usually think that they have gone wrong somewhere. To avoid problems, just tell them you're going out with Todd. Now that isn't a lie because you are going out with him—but you don't have to say what you're doing."

Taking part in these conversations made David feel stronger and more secure in his gay identity. He accepted the fact that, for the time being, he was in hiding, but he felt less frantic about pretending to be a heterosexual when he knew he could reveal his "real" self on the computer.

David continued in the support group for several months. He came out to his parents just before he left for college. They were stunned by the news, but they were able to absorb it better than they would have been able to when David was younger. David told his best friends, too. He had decided that no matter how they reacted, he could no longer live behind a mask. His computer friends knew that he planned to make the move, and they waited anxiously for word from him. Here's the message they received:

"I must share something wonderful that happened to me. I 're-introduced' myself to two of my very best friends—one a guy and one a girl. They took the news of me being gay very well.

"I was so terrified at what could have happened, but everything turned out great! They were so accepting. We seem to be a lot more open with each other now about everything. And they are both being very supportive.

"I just wanted to let the world (part of it) know how good I feel about what I did. And all of you guys helped me to do it. Thanks all! I'll always love you, David "

"I have an electronic diary"

It's 1 A.M. when Jan, a lively 16-year-old, returns home from a "special" date with Chris, a boy she's been dating for four months. Jan has ambivalent feelings about Chris. She likes his looks and his intelligence, but sometimes he scares her. He can lose his temper at the drop of a hat. Tonight, for example, he nearly hauled off and socked the man in front of him at the movies, just because he politely asked Chris to stop talking.

Jan is deep in thought when her mother knocks at the door. "I thought you were going to come right home as soon as the movie was over," she says. "It must have ended well over an hour ago." By the time she finishes her sentence, her voice has risen a number of decibels. Jan knows better than to start a conversation now. She mumbles something about trouble with the car and says good night.

As soon as her mother is gone, Jan switches on the computer. Jan is the computer whiz in the family, and the machine is her special province. She logs on to HSX's support group for teenagers and spills out her feelings about the date. "Well, tonight it finally happened. Chris handed me the big ultimatum: Go to bed with him or we're finished. He's tired, he says, of being frustrated. He claims he loves me, but he just can't stand it any more.

"Physically, I sure do feel like giving him what he wants. But I just don't know. Part of me doesn't trust him. Oh, I know he'd use a condom and all that. What I mean is—I don't trust him emotionally. Sometimes I think he isn't all that grown-up." Jan goes on to describe the incident in the movie theater. Then she concludes, "Chris can be weird, but I sure don't want to lose him. I'd miss him like anything. Wow—I wish I could make up my mind. Any suggestions, folks?"

Jan's been using the computer like this, as a diary, for the past year. Everyone who's ever been an adolescent knows how powerful feelings can be at that time. Lots of teenage girls get rid of excess emotions by confiding them to a diary and locking it with a key. Then they hide the book away in a drawer.

Jan has a diary that talks back to her. Whenever she comes home from a date, she logs on to the computer and records her feelings. The computer knows which boys turned her on and which boys were duds. It knows what clothes she decided to wear and what music the rock band played at the dance. It knows her disappointments and her triumphs.

The only difference between the computer and a real diary is that it's far from private. Rather than locking her thoughts away, Jan shares them with computer friends. This way she gets feedback from her feelings.

By noon the next day, for example, two messages appear on the screen in response to Jan's. One is from Emily, a computer pal who's in her mid-forties. Emily has already heard quite a bit from Jan about Chris. In fact, she's had uneasy feelings about him for several weeks. "When I was in college I dated a boy like Chris," Emily's message says. "To tell you the truth, he was my first sexual experience. He was also my first experience in getting a black eye. So watch out, Jan."

A message from 18-year-old Herb says, "Jan: If Chris really cared about your feelings, he wouldn't be putting all this pressure on you. Also, why don't you level with him about how upset you get when he blows his stack? If you're afraid to tell him how you feel, you don't have a good relationship, I'd say. Well, I'm just saying it because you asked."

Jan replies: "Thanks, you guys. I guess that Chris makes me even more nervous than I've been letting on because I haven't been telling him what bothers me. Now, I've made up my mind to have a long, honest talk with him. If you don't hear from me again, you'll know I've got a knife in my back. Ha, ha. Hugs—Jan."

By using the computer as a diary, Jan can both communicate and have privacy at the same time. There are some things she doesn't want to share with her real-life friends or her mother, who, she says, sometimes reminds her of a dragon. On the other hand, she does want to share these experiences with someone—and her computer friends fit the bill. They're far enough away for her to feel safe, yet close enough, via the magic of electronics, for her to make contact with them. Some of those people are adults, so they can give Jan the perspective she might have gotten from her mother, if she trusted her mother enough.

Someday Jan won't need a computer diary to sort out her feelings. She'll be more secure about her own judgment. Until that time arrives, it feels good to know that the computer is always a willing confidante.

The Diaper Pail Fraternity

Alex has a fetish that's relatively uncommon. He's an infantilist, who gets sexual pleasure out of childhood objects.

Like many other infantilists, Alex enjoys masturbating while wearing adult-size diapers. In Alex's view, he's a 31-year-old man on the outside, but a two-year-old on the inside. By returning to infancy, he can dismiss the insecurity of adult life. His sexuality is dominated by childhood experiences and emotions.

Since Alex realizes that his infantilism is little understood, he keeps it a secret from everyone but his wife. Although he argues that his life-style is unusual but not abnormal, his diaper fetish has always made him feel lonely and isolated. He's not even sure that his wife approves, although she goes along.

Lately, some of Alex's loneliness has abated. He has been able to share his fantasies with other infantilists in a special HSX support group for alternative sexual styles. The infantilists, some of whom belong to an organization called the Diaper Pail Fraternity, share more than their feelings; they also share practical information, such as where to obtain mail-order catalogs for baby supplies.

All Alex asks of the world is not to be judged harshly. In his view, his infantilism represents a return to a ''magic land''—the world of babyhood where he would rather dwell. One night, Alex left this poem on the computer. It explores the inner world of ''diaper babies'':

If only you would take my hand
I could show you my magic land
A place of eternal springtime
This is someplace uniquely mine

My Mommy's lap is always there
Safe and warm to banish fear
And when I fall and scrape my knee
Someone always comforts me

My teddy bear and I can play
Till darkness overtakes the day
When bedtime comes I know I'll see
A baby's crib awaiting me

My Daddy pins my diaper on
I know he loves his little one
Then on his lap with Dr. Seuss
Nursery rhymes and Mother Goose

I suck my thumb and close my eyes
Daddy looks at me and smiles
Softly kissing my forehead
He gently tucks me into bed

And if I waken in the night
Cold and wet and filled with fright
My Mommy is there to ease my fears
And wipe away my little tears

She brings my bottle of warm milk
And with a touch as soft as silk
My diaper is changed and soon I'm dry
While Mommy hums a lullaby
My magic land I know I'll keep
As Mommy rocks me back to sleep

Now if my hand you cannot take
And my land you would forsake
Judge not so harshly such as I
For fear my magic land would die

After reading this poem, 27-year-old Simon was inspired to share his own experience:

"I am a diaper fetishist and have been one as long as I can remember. Even when I was four years old, I was infatuated by diapers and baby clothes. At age twelve, I started wearing diapers and rubber pants in secret. At age twenty, I graduated from college and moved to a big city where no one knew me. No one at work knows that I wear diapers there and wherever I go.

"I make a lot of my own clothes, such as pants with snaps up the legs, just like a baby's. I also make my own bibs, and I buy rubber pants and diaper pins at a drugstore. I make my diapers by sewing five Curity diapers together. No one in the world knows about the dozens of diapers that I own, not to mention the rubber pants, baby powder, baby bottles, baby food, pacifiers, fold-up crib, and the high chair I keep locked away in the closet. No one knows what it's like to have to hide everything away carefully so that when a friend drops in, your secret will not come out."

A number of non-infantilists also read Alex's poem and were touched by it. They felt it increased their comprehension of a life-style that's quite different from their own. "In all honesty," Karen said to Alex, "before I got here, I had never heard of diaper-wearers. I still don't understand it, but I'm glad to have learned something about this because it has made me realize that I can accept something without understanding it. Maybe that's what the computer is all about, accepting people's differences."

"I fell in love with a mystery woman"

Ricky was the kind of guy who was better at meeting girls on the computer than in real life. He was cerebral and shy. Unless someone approached him, he'd fade into the background at social events.

Ricky had no trouble making friends on the computer, however. There, he could quickly reveal his sharp intellect. Ricky found it easier to relate to people when they were disembodied. That way, they were less threatening to him.

Still, Ricky longed to meet an intelligent girl on the computer and get to know her in real life as well. At 26, he was ready for a committed relationship.

Almost as soon as Lauren logged on, Ricky was attracted to her. He liked her incisive comments and her lively personality. She was not only smart, she was clearly outgoing—and Ricky admired people who were more assertive than he was.

Lauren said that she lived on the West Coast. Her folks were well off, and she had a job as a fashion model, "just to pick up some extra cash." She drove a Jaguar, she loved to shop, and she also liked to read a lot. In addition to reading, she shared two other passions with Ricky—chess and trivia games.

Ricky tested Lauren on trivia. "Who's the only president actually buried in Washington, D.C.?" he asked. "Woodrow Wilson," came Lauren's quick reply.

Lauren challenged Ricky to a game of chess. "Let's see how smart you really are," she said. After they had commanded their computers to "/talk," they had privacy for the chess game. They proved to be evenly matched, and the game went on for weeks.

Ricky found himself lying awake at night, thinking about the move he would make the next day. He also found himself fantasizing about Lauren. What did she look like? Whenever he asked, she was evasive. She must be good-looking, he thought, because she was a model.

Finally, Lauren came up with a description of herself. She was tall and blonde, with green eyes—although the eye color came from tinted contact lenses. Her real eyes, she said, were blue. Ricky gave her the particulars of his appearance. "I'm nothing great to look at," he said. "You sound terrific to me," Lauren replied.

Two or three chess games later, Ricky and Lauren had become fast friends. They played other games on the computer as well, solving puzzles and cerebral quizzes that one or the other of them invented.

More and more, Ricky was possessed by the thought of Lauren. She was captivating him, not only with her intellectual ability, but with her sexual provocativeness as well. Suddenly, in the middle of a chess game, she'd demand to know what he liked sexually. "I'm crazy about oral sex," she once commented. "If you win the next game, I'll do it to you. But first, you've got to make the right move."

When Ricky did "make the right move," Lauren rewarded him with a graphic description of how she was orally stimulating him. Increasingly, their game-playing became both sexual and cerebral —a combination that was irresistible to Ricky. He even found her brusque use of sexual slang sophisticated rather than crude. He saw Lauren as a woman of the world.

Ricky was in love with Lauren, and he wanted to get to know her in real life. She was, he thought, the woman of his dreams, a woman who could stimulate him sexually as well as mentally. He had to meet her.

But when Ricky asked Lauren if she would "voice" with him, she said she couldn't "for personal reasons." Ricky continued to press her and she finally replied: "Three years ago, I had squamous cell carcinoma. To remove the tumors, they had to perform a partial laryngectomy, so I lost about 50 percent of my vocal cords. This has left me with a whispery voice. With an amplifier device, I can converse on the phone; however, it causes me embarrassment when I first talk to people. I can pretend that I have a cold for only so long."

Ricky was overwhelmed with pity for Lauren. No wonder she enjoyed being on-line so much. She could carry on conversations without having to feel handicapped. Still, no matter what kind of voice she had, he longed to hear it. "I want to talk with you," Ricky answered, "I love you."

"Please don't call. I'd be too ashamed."

But Ricky was determined to make contact. He knew Lauren's last name, but when he called information in her small California town, there was no listing under that name. "The phone is in my stepfather's name," Lauren explained. "If you call, he'll find out what I've been using the computer for. He'll be furious because he pays the bills. That's why I never make calls from here."

By this time, several of Ricky's computer friends were becoming concerned about him. They knew how much he cared for Lauren and, to them, her story just wasn't plausible. Flo, who was a nurse, pointed out that laryngeal cancer mostly struck middle-aged or older men who had been heavy drinkers or smokers. "That stepfather may exist," Flo said to Ricky, "but it's also possible that I'm the Pope. Why don't you ask Lauren for her driver's license so you can check it out? *That* surely can't be in her stepfather's name."

When Ricky raised Flo's suggestion with Lauren, she became furious. "You're just too mistrustful and paranoid for me," she said. "It's a good thing we didn't develop a serious relationship. I needed a little understanding, and all I got was suspicion." With that parting shot, Lauren disappeared from the service, never to be heard from again.

For a long time, Ricky was angry at his friends—and even more angry at himself for driving Lauren away. He refused to accept his friends' contention that Lauren was a "cross-typer"—a man pretending that he was a female for a turn-on.

Cross-typers, though far from typical among HSX subscribers, can be a real problem because they're difficult to detect. In Lauren's case, however, there were some obvious danger signals. Generally, it's wise to be wary of any person who claims to have a super-glamorous background—the modeling job plus the Jaguar plus the wealth, for example, should have made an alarm go off in Ricky's head. Two other reasons for suspicion: Lauren's refusal to make voice contact and her claim to have an unusual disease.

Of course, Ricky had good psychological reasons to trust Lauren. It wasn't easy for him to meet women ho responded to his needs. Lauren played to his fantasies perfectly—perhaps too perfectly.

Even today, though Ricky is once again on good terms with his friends, he refuses to admit that Lauren was not the woman she claimed to be. Someday, he's sure, she'll return to the computer and give him a reasonable explanation for her behavior. Then, they'll be able to resume their relationship.

ELECTRONIC CONVERSATION

"Why do you use HSX?"

Evan: I wonder why people decide to use HSX. Why do they come here, as opposed to spending time in "real life" for example?

Debby: One of the reasons I spend time on-line is that I have little time for social activities. I'm a full-time student and I also work part-time. So, rather than going through the hassle of scheduling activities with my friends, it's a lot less time-consuming to flick on the computer and be in a conversation in a matter of seconds.

Robert: The reason I'm here is that it's easy to make significant friendships this way. Regardless of the clowns that appear now and again, the general populace of HSX consists of nice, real people who care about others in their triumphs and defeats.

Debby: One of the things I like about HSX is that it isn't "real life." Here, we can escape from one aspect of real-life relating, accountability. If we don't click with a person, we can always tune out. In real life, it's not so easy. Sometimes we have to stay around and work it out. Or endure the pain of a broken relationship for some time.

Al: I'm here because I find a real welcome and a good time, as my computer bills will attest! For me, it's like a hobby or a harmless vice, as I often have to remind my patient and long-suffering wife.

After all, I don't belong to the country club or have any other major indulgences.

Evan: Well, I joined HSX to establish relationships. I'm finding that the caliber of people I meet here is the highest. I can meet people who I wouldn't run into otherwise. I talk to many people who are involved in exciting careers and they share their thoughts and feelings with me. I love it! Where else would I have the opportunity to meet so many smart people—from all over the country—in one place?

Leonard: The reason I chose HSX is that I like to play. This is like a big playground for me. I adore it when people pretend to be serving ice cream or blueberries to each other. Once, I was even in an imaginary crap game. I've had a number of compusex experiences, too. One of them ended with the woman sending me "pretend" flowers. Of course, I've never received a "pretend" bomb, but maybe I will someday—if I deserve it. The adventure is why I like being on HSX.

 *"Computer friends are the perfect friends"*

Oliver: The reason I like meeting people on the computer so much is that we can be honest with each other, whereas in real life, one can't always be. I might add, too, that sex plays a small part in meeting people here. I go to bed with very few of my computer friends. In fact, I've made voice contact with very few of them. Yet, after I turn off the computer, I'm lonely for them.

Anne: I can open up more with strangers on a keyboard. Don't get me wrong. I do have a small circle of personal friends, but I enjoy myself more when I'm either with my kids, animals, or the computer. I just find that relationships here—whether they be friendships or compusexual relationships—are much more intense.

Bill: Do you think it's healthy to be so attached to the computer friends we make?

Anne: Yes, I do. Because computer friendships can be actualized, if you want them to be. In fact, one of the friends that I made online is sitting here beside me right now. I think that getting to meet him and actualizing my friendship with him and others has been an important part of letting me become less lonely. After all, I get to find out that those letters on the screen are connected to real people—and that people are as nice as they seem.

Bill: Well, I think that computer friends are less real. Sometimes the people I meet here just seem like a bunch of words. Once I'm off the computer, my real life takes over.

Connie: The thing I like about my computer friends is that talking to them can calm me down. If I've had a bad day, I can let off steam on the computer. No matter how rotten the day was, I know that I can share it with somebody. The keyboard is so impersonal—I can open up to it without fear of being rejected. In fact, when I gave birth to my baby, I spent the first part of my labor on the computer getting support from friends. It made the whole thing go by more quickly.

Oliver: For the most part, computer friends are the perfect friends. They don't judge you, question you, or out-and-out call you a liar. It's a calm, accepting friendship, even though we may never meet.

Connie: My point exactly. Where else can you get that kind of acceptance? Can you walk into any other new surroundings—a bar, for example—and get the instant response that you get here?

JoAnne: I can relate to what Oliver said, too. I have many friends here on the system. I find that even with the ones I have met or voiced with, there's less tension and more closeness than in other relationships I've experienced.

Oliver: I've found that the very knowledge that the computer is there, that I can go on it when I feel lonely and find someone to talk to, makes my life less lonely altogether.

Bill: Are you sure you're not just escaping from reality? I truly enjoy the company I find here. But there are also times when reality sets in, and I have to live a true life with flesh-and-blood people. I tend to relate better to people that I can meet in the flesh.

JoAnne: It would be nice to personally meet everyone I talk to here. But I still consider on-line conversations to be part of my reality, even if I may never see the people in person.

Connie: Bill, why do you have to separate your computer friends from real-life friends? Isn't being on the computer somewhat like the movies, where you know an actor or actress but only through fantasy? You think you know the feel and touch of the person, but you'll never meet them. Here too, as much as I like the people, I'll never know their true reality.

Edgar: I've been listening all along and here's my two cents. Computer relationships are real because computers are forms of communication as well as tools. They communicate an individual's personality, but in a different way. It doesn't really matter how it works—it just works for us. Being on the computer makes lonely people less lonely. They've found a new way to relate. And that's enough to satisfy me.

"What are the responsibilities of compusex?"

Mark: Tonight we'll be talking about computer relationships, compusex, and compulove. What kinds of responsibilities are involved in on-line relationships?

Henry: The main one, I think, is to be honest. For example, I ran into a wonderful person on the computer. We exchanged electronic mail for quite a while. She described herself as a ''Cinderella.'' That turned out to mean that she was too short for her weight, like 275 pounds and five feet, five inches tall. What a surprise!

Laura: You're right. A primary responsibility is to be honest about yourself. Especially your sex!

Mark: Would you care to elaborate on that comment?

Laura: Yes. A woman once represented herself to me as male and let it drag on much longer than was necessary. I've avoided those encounters ever since.

Toby: Well, lots of people get on-line to escape. I know that I do. Everything here is unreal. I do try to be honest, but I take everything and everybody I meet here with a large grain of salt.

Mark: Some of us here take everything quite seriously. What's to be gained by fibbing, even a little?

Toby: Escape. This system gives people an opportunity to misrepresent themselves and some of them just can't resist. You can be anything you want to be here. It's a fantasy medium.

Henry: You can say that again. Just last night, some woman—at least I think it was one—asked me to ''/talk'' with her. When we were alone, she wanted to know if I could fix her up by giving her a massage. Now, there's no way you can do that over the computer, so I terminated the conversation.

Gloria: You're wrong. Anything is possible. You just have to have a lot of imagination. That's what compusex is all about.

Laura: Compusex is unusual, but not impossible. Other women have told me that they can achieve orgasm on-line. I asked one woman how she does it and she said: "With one hand—while I'm typing with the other—and it takes practice."

Mark: Well, if you're on-line hunting for impersonal sex, most likely you don't mind if there's a liar, impostor, poseur, or actor on the other end. If you want to get yourself off, by golly, you'll do it either on-line or later. You're just having a little foreplay with yourself. Of course, if you're interested in a serious relationship, dealing with an impostor presents a problem.

Henry: Well, I come here to meet other people and to make friends, albeit electronically, not to be blatantly lied to. I think we owe it to each other to be at least close to reality.

Gloria: No, reality is not one of the rules.

Mark: I feel that if you can't be honest about who you are on the screen, the only person you're lying to is yourself.

Henry: Gloria, how can you say that reality is unimportant? How can one be intimate with a liar? Sex, even on-line, is dishonest if just one party is getting it off.

Gloria: But how honest is the real world? Aren't you lied to all the time by car salesmen, real estate agents, people that work in personnel offices?

Toby: As I said before, this is a fantasy medium—and I don't think that compusex involves any responsibilities.

Mark: You're wrong. This is not a video game. There are real people on the other end of the line. In fact, most people I know take computer relationships quite seriously. I know of at least three weddings that developed from relationships that started on-line.

Henry: I agree. This is not fantasy. There are real people out there. They fall in love, they get angry, they get hurt. If both people treat a relationship as fantasy, well, okay. But that's not usually the case. I think you have just as much responsibility here as you have in the real world. If you start any kind of relationship—on-line or off—you should be treating it honestly.

Mark: I wish that someone would write a computer program that could weed out impostors. But in the meantime, I think we have to treat compusex as we would sex in real life. We should try to screen out the phonies before we get involved with them. It's hard to do that here, but it's worthwhile to make the effort. I know one great way to expose a man who's posing as a woman. Ask him to explain what a "petite" dress size is. Very few men know the answer to that one. So if "she" can't tell you the answer, break contact. And better luck in your next relationship!

"Am I addicted?"

The computer poses a new problem: addiction. Some people become so dependent on their computer relationships that those relationships become more important than friends—and even family. Listen to a group of addicts discuss why they've become addicted and how to deal with the problem.

Phyllis: I really prefer the "me" I am on the computer, and I enjoy the people I know on-line more than almost any I know in the off-line world.

My husband and children complain bitterly that I spend too much time on the computer. Right now, for example, they're at the movies, and I'd rather stay home and be on the modem. They say I'm "addicted" and "negligent." I'm worried that they're right, but I enjoy the computer too much to give it up. Has anyone else got this problem?

Gina: I do. A few weeks ago I was vacationing at a friend's beach house, and I found myself truly upset because I couldn't log on. I was agitated, had sweaty palms, and kept racing around the house and talking to myself. Am I doomed to live the rest of my life talking through my keyboard? What does this mean?

Rex: Gina, for my money (and I spend lots of it here) you're definitely addicted. However, you're not the only one. I get many of the same symptoms you mentioned if I don't get on-line at least every other day. I don't think it's bad. It's just expensive and occasionally inconvenient. In the long run, though, it's worth it.

Phyllis: I don't know if it's worth it. My husband goes crazy when he sees the bills. I wish I could learn to regulate my binges. Any ideas?

Richard: The best way to wean addicts away from anything is also the harshest: cold turkey, pure and simple. That's how I quit smoking. Here's an idea: Get rid of the computer. Or, enforce a time limit. Put the computer on a twenty-four hour timer that only sends power to it for thirty minutes a day. Put the timer in a strong box with a lock and throw away the key.

On-line relationships are sure attractive, but I find that they don't beat waking up beside a cuddly body in the morning.

Angel: I sympathize with Phyllis. My computer is in my bedroom. I get many complaints from my family that I disappear there every night after dinner. If the "live" people around me were more interesting, I would be with them. Maybe Phyllis has the same problem. Faced with the choice between TV, a movie, or an on-line conversation, I choose conversation.

But I do know a number of addicts who have jeopardized their lives by running up computer bills. One friend gave me her password, and I would change it when she used up her "budget" for the month. That worked, but I used to get phone calls from her pleading for "just a few more minutes."

Getting love and understanding from intelligent, warm people is addicting. I can even understand why people enjoy compusex. I, for one, need all the love and understanding I can get. People here have been wonderful to me. No wonder I'm addicted to them.

By the way, it isn't the computer that's addicting. It's the communication with other people. That can't be bad for you.

Richard: I don't know. Computer communication is different. It's more like the radio dramas of my youth. You fill in a lot of details about the other person with your imagination. Now imagination is great. But when fantasy becomes more important than real life, there's a problem.

Phyllis: If all you people are a fantasy, I hope I never stop fantasizing. Giving up smoking would be a piece of cake compared to giving up my computer!

I wish my husband wouldn't be so jealous of you, though. I think that's why he says I'm an addict. I wish I could convince him that I'm not out to "bust up" our marriage, but simply to have fun talking, talking, and talking to my friends.

"My feelings about HSX"

Sally: Human behavior is just as varied over this medium as it is in the "real" world. We frequent users always seem to make that distinction—between our world of telecommunicating and the other parts of our lives. But I find that what goes on here is just as real. I have seen and felt a wide range of emotions taking place—from love to hate and everything in between.

I've had more than one person tell me that they have discovered emotions that they hadn't experienced since their high school days. Now, most of these people are fairly new to the system, so I don't know if, as time progresses, they go through a "maturing" process. Maybe that is why there is such a high turnover of users. Maybe a type of maturation occurs, and this medium is no longer useful for experiencing those youthful feelings.

I can see this transformation in reading some of the messages on the board. A message starts out as one "seed" for thought, and then each writer adds a little of himself. There are times I find myself getting so involved talking with one person, that when I look up at the subject, I find that we are "supposed" to be talking about something entirely different. It's nice when I can see that both of us have experienced growth during this dialogue.

I have seen love, warmth, compassion, anger, and maybe even hate in these messages. But in each case, the writers have made a thoughtful contribution. They will say not only what is on their mind, but what is in their heart. Those inner feelings that have been suppressed for so long are "exposed" to the public. And it is nice to find a public so understanding and caring. We are all growing as individuals, but we are doing it together.

Reflections on computer friends

Arthur: It has been nearly eight months that I have been a regular here on HSX. In that time, I have made dozens of acquaintances, and several very close friends. I have discussed, in conference and on the message board, every subject under the sun. I have met a wide range of people from all walks and stages of life. I have asked and answered questions that I would never think of discussing in "real life," and in doing so have found friends who are very different from me and some who are very similar.

But in all the hours of interacting with people in this unique medium, I have come to one conclusion: Behind all the jokes, the teasing, the banter, and the fun we have with one another, all of us have experienced hurts, disappointments, and the hard knocks of life, which makes us very much in need of true friends. We have sorrows that we don't talk about (except in quiet moments), we have shattered dreams that haunt us, and we all have the deep need to be accepted and needed by others.

Behind every "hiya" is a complex person who is fighting the battles of life just as we are. Behind every "howdy" is a person who often feels overwhelmed and is sometimes lonely. Behind every "bye" is a person who is coping with the pain of unfair treatment.

But each person has dreams that want to get out. There are happy memories wanting to be shared. There is love that wants to be given. There is hope that tomorrow will be a better day. There is a deep desire to trust and be trusted, and in doing so, to build enriching new friendships, either contained within the computer or extending off of it.

As we all roam the halls of HSX, I hope that we will not forget the depth of the needs behind those words we see on the screen. And should we forget, we just have to look in the mirror. The person on the other end of the line is not much different from us.

QUESTIONS AND ANSWERS

If I use my diaphragm and contraceptive jelly, is it safe (from pregnancy, that is) to make love with my husband in our hot tub or in a swimming pool?

It's not advisable.

When a woman bathes or swims, water usually doesn't enter the vagina because its soft walls fold into each other. But when you have intercourse while immersed in water, the thrusting penis can pump water into the vagina. That could dislodge the diaphragm, ending its effectiveness in preventing pregnancy.

My wife doesn't like to perform oral sex on me. She says the taste, even before I ejaculate, makes her sick to her stomach. Is this possible, or is it a psychological hang-up? Can I change the taste through diet?

You can lighten up this situation by focusing not on ejaculation, but on having fun beforehand, recommend sex therapists Richard Reznichek, M.D., and Carol Reznichek, R.N., M.S.N. For starters, try having oral sex in the shower with the water running. Your wife may be more comfortable there. Be sure to reward and reinforce her efforts with one of her favorite ways to finish!

Diet won't change the "taste" your wife experiences. But have you considered applying some flavors to your penis that she might enjoy? Maple syrup? Whipped cream? Chocolate sauce? Be patient and work inventive oral sex into your love play little by little. Remember that your wife is unlikely to enjoy or accept anything that she doesn't have time to become comfortable with.

I'm 23. My 19-year-old girlfriend and I have decided not to have intercourse until we're either engaged or married. Instead, we masturbate each other to orgasm.

I have no problem bringing her to orgasm. But she has not been able to do the same for me. When I tried to show her how to grip me more forcefully, she refused. Her former boyfriend, she claims, had no trouble reaching orgasm.

Could the fact that I masturbate frequently have made me less sensitive to her touch? Am I the problem or is she?

As we see it, the main problem is your girlfriend's unwillingness to discuss the sexual difficulty with you. Open communication between sex partners is very important. Consulting Editors Armando DeMoya, M.D., and Dorothy DeMoya, R.N., M.S.N., the first sex therapy team to be trained by noted sex researchers William Masters and Virginia Johnson, point out that "communication is the cornerstone of effective and satisfying sexual functioning."

Don't worry that masturbating is damaging your sex life with your girlfriend. The opposite is true. Masturbation has helped you learn what feels best to you and how much stimulation you require to reach orgasm.

Your challenge is to convey what you like in a way that will not make your girlfriend feel inadequate and defensive. Her frustration has most likely made her feel like a sexual failure. Feeling defensive, she protects herself by telling you that there's nothing wrong with her (after all, her previous boyfriend was satisfied with her technique).

For your part, her resistance to doing what pleases you must be frustrating. That can contribute to your difficulty in ejaculating.

Explain to your girlfriend that people have different sexual responses and preferences. What was fine for her previous boyfriend is obviously not suited to you. Be sure to make your suggestions constructive so that she doesn't take them as criticism. For example, instead of saying, "That's not fast enough," you can say, "That's very good. A little faster would be better." Or, "It would be wonderful a little higher up."

If she's receptive, you might put your hand over hers to indicate the location, grip, and speed that are most pleasurable for you.

I heard that if you use birth control pills, you're somewhat protected against VD.
Is this true?

On the contrary. Taking the birth control pill (oral contraceptive) may make a woman more susceptible to sexually transmitted disease.

Some evidence indicates that the Pill creates an environment in the vagina that encourages the growth of gonorrhea-causing organisms.

Other birth control methods, by contrast, offer some protection against venereal diseases. For example, the condom—used in conjunction with a vaginal spermicide—may help protect against gonorrhea, nongonococcal urethritis, trichomoniasis, and syphilis.

I'm a 28-year-old man.
Back in junior high school, I had a "crush" on another boy. For one year, I tried to spend as much time as I could with him. I'm definitely not homosexual, but this little episode has always bothered me. Should I be worried?

What you experienced was a same-sex crush, a very common experience among both boys and girls during adolescence.

Puberty unleashes strong feelings of sexual attraction and affection that demand expression. These feelings often combine in a crush—an intense, almost obsessive love for a special person. Professionals who have studied the phenomenon of same-sex crushes

CONFESSIONAL BOX
<u>"I have to go through the pain all over again"</u>

<u>Lane, (39)</u>: I've gone through two major break-ups in the past ten years. First my wife left me, and recently the woman I had been living with did the same thing. I miss her and her two sons. I still love her much more than I loved my wife, but ever since we broke up, the pain of the first break-up has returned.

I find that it's impossible to avoid the pain. The best I can do is to let the pain happen and work it through. Maybe then I can find some happiness—but I doubt it.

You see, I'm stuck with the reality of being 39 and trying to start all over again. I've forgotten how to date. My wife and I had no children, and now it looks as if that hope for me is gone. I think I'm too old to try raising children. Or is it still possible? I'd love to have some so badly.

consider them a normal part of maturing. Indeed, crushes can help a youngster develop the capacity to love and be loved.

Teenage infatuation, whether with the same or the opposite sex, is generally beneficial, say psychiatrists Eugene A. Kaplan of the Upstate Medical Center in Syracuse, New York, and Frank A. Johnson of the University of California in San Francisco. In their view, a crush represents a healthful movement toward feelings of love and admiration for another person. "For many adolescents, the crush is a common transitional phase between the self-centered preoccupation of the child and the other-directed sensitivities of the young adult."

Crushes help adolescents learn about who they are and what they want to be. Many adolescents adopt the characteristics of the loved one, trying on different roles and exploring different interests. This can help them define their own personalities.

Because same-sex crushes among boys are a little-discussed reality, a boy who feels a special warmth for another of his own sex may panic. He may think the feeling signifies that he's gay. One result of such fears is that boys typically develop friendships that are less intimate than those between girls.

A boy is especially likely to fear that he's homosexual if the crush has led to sex play. But such play is a fairly common occurrence among boys with mutual crushes.

Comments Dr. Lester Kirkendall, a sociologist with a specialty in teenage sexuality: "In general, persons with affection for each other want to touch and embrace one another—to have close bodily contact. Little wonder then that two adolescent boys, close pals, have physical contacts which lead to genital explorations and sexual associations. Such experiences satisfy both their desire to be close and their curiosity, and also give them physical pleasure."

Most crushes are of short duration. They rarely survive more than a few months, hardly ever longer than a school year. When they start to fade, crushes usually die quickly, often with a final burst of self-doubt, jealousy, even hatred. But professionals believe that even this agony can be beneficial. The end of a crush can help adolescents develop the strength to cope with disappointments.

Note psychiatrists Kaplan and Johnson: "The process of maturing requires that one develop the capacity to sustain and tolerate loss, rejection, and the nonreciprocity of some human relationships. Such griefs . . . painful though they may be, are part of the human condition and help to shape character."

How can I find and stimulate my girlfriend's G spot?

Preliminary research suggests that the Grafenberg spot, or G spot, a female genitourinary structure, may be an erogenous area that's present in all women.

According to findings reported in the *Journal of Sex Research*, the G spot is located deep within the upper front wall of the vagina, adjacent to the urethra. The G spot seems to be an oval lump about the size of a lima bean. It reportedly swells when stimulated.

Considerable pressure is needed to stimulate the G spot. Usually, the stimulation is done with the partner's fingers. That's because it's difficult for a woman to reach the G spot by herself when lying down.

Evidently, many women can reach orgasm through stimulation of the Grafenberg spot. During intercourse, the G spot is most likely to be stimulated by the penis if the woman is on top.

Help! Every time I make love with my girlfriend (I'm 20 and she's 18), the condom breaks after one or two minutes of use. I've used different brands, with the same results. Once, I even put on two condoms at the same time, but I couldn't feel a thing.

I am very worried about making love and bursting the condom when I climax. My girlfriend doesn't want to go on the Pill because if her parents found out, all hell would break loose. Please give me a solution, fast.

First, consider whether you're doing anything that might cause the condoms to deteriorate and break.

Condoms are likely to dry and crack if they're exposed to heat. So don't keep them in your wallet (body heat can speed deterioration) or in the glove compartment of your car, or in any other unusually warm place. Don't use condoms that are more than two years old. (Check the date of manufacture, especially if you bought them from a vending machine.)

Lubricating condoms with vaseline or cold cream can cause deterioration. So you're best off using prelubricated condoms. They're less likely to break than unlubricated ones.

For added protection, your girlfriend can use a spermicidal preparation (such as foam) at the same time as you are using a condom. The combination of a condom and foam is very effective contraception—just about matching the effectiveness of the birth control pill. Like condoms, spermicides can be bought without a prescription at pharmacies. But it's not a good idea to give up on condoms and use only spermicides. Alone, they're not very good protection.

If a condom breaks while you're having intercourse—and if your girlfriend is not already protected with a spermicide—she should immediately fill her vagina with a spermicidal foam, jelly, or tablet. She should leave the tablet in place until it dissolves.

She should not douche. Douching can drive sperm farther up toward the uterus and cause pregnancy.

CONFESSIONAL BOX:

"I don't want any more casual sex"

<u>Greg, (30)</u>: I've become celibate for religious, health, and emotional reasons. I spent most of my life following the crowd, in pursuit of casual sex. In my maturity, I've found that getting laid without commitment is just that, "getting laid." It's not emotionally satisfying nor is it healthy, as evidenced by today's epidemics of sexually transmitted plagues. I use the term "plagues" because I believe this is the way God wants us to view them— as punishments for the wrongdoing of "the sexual rev- olution."

I think I've O.D.'d on sex. Ever since I was in college (I'm 34 now), I've had all the sex I could want.

Sometimes I was sleeping with three or four girls at a time, and even engaging in mate-swapping and group sex. None of these relationships lasted long, but it didn't seem to matter, because there were always more women around.

Recently, this whole way of life has gone sour. Sex is no longer the kick it used to be. I've decided not to have sex again unless it's with a woman I want to marry—or at least care about very much.

I haven't had sex for four months now. Is this weird, or what? Is it dangerous to my health? Is it normal?

Dr. Joel Moskowitz has coined the term ''secondary virginity'' to describe what you're experiencing—a reaction to the high psychic cost of casual sex.

Dr. Moskowitz, a psychiatrist who has been director of clinical services at the Resthaven Psychiatric Hospital and Community Center in Los Angeles, first observed ''secondary virginity'' among college students and young adults. But it's also known among people your age.

A growing number of young people—after finding sexual experimentation hurtful, disillusioning, or emotionally empty—become scrupulously chaste. Like you, they often vow to save further sex for marriage, or at least for an emotionally close and stable relationship. The decision to abstain is closely linked to a desire for commitment as part of the sexual experience.

Like some others who give up sex for a time, you may find the experience instructive and satisfying. Often there is a sense of freedom in not feeling compelled to engage in sexual activity every time the opportunity arises.

You needn't worry that abstinence will be damaging to your health. Lack of sex causes no organic illness or medical condition. Nor does abstinence cause psychological disturbances, as is popularly believed.

My wife is able to have an orgasm only with manual stimulation of her clitoris. Did God put her clitoris in the wrong place?

It's most unlikely that there's anything wrong with the location of your wife's clitoris.

The clitoris, a small knob of tissue located above the opening of the urethra, is surrounded by a fold of tissue, the clitoral hood. The hood is attached to the labia minora, the liplike structures at the entrance to the vagina.

Direct or indirect stimulation of the clitoris usually brings about orgasm in the female. During intercourse, the thrusting of the penis moves the labia minora down at the entrance of the vagina. This rhythmic motion can stimulate the exquisitely sensitive head of the clitoris.

However, the action of the penis alone is usually not enough to lead to an orgasm. Most women require additional stimulation, as with a finger, to come to orgasm during intercourse. Surveys find that about 70 percent of women do not experience orgasm through intercourse alone. Thus, the additional stimulation of the clitoris your wife requires for orgasm is actually the way most women come to orgasm.

Many men, however, have the expectation that women ought to have an orgasm through penile thrusting alone—as most men do. They convey this expectation to their female partners, who come to feel that manually or orally induced orgasms are inferior. Women may sometimes feel that there is something wrong with their sexual anatomy.

Recent research has shown that women have no reason to come to such conclusions. Give your wife the manual stimulation she needs. Her clitoris is in the right place.

I've heard that if a man blows air into the vagina while performing oral sex, it can cause an embolism, and possibly death.

But when I engage in normal intercourse with my fiancée, some air gets into her vagina. Can this be fatal? If it's dangerous, what can we do to keep it from happening?

You're probably referring to the rare deaths of pregnant women that have occurred when air was blown into the vagina.

Bubbles of air that are blown forcefully into the vagina can pass through the vagina and the open cervix and into the uterus. Once there, the bubbles may get into the dilated uterine blood vessels that surround the fetus. From there, they may travel through the woman's bloodstream, where they can block circulation in the lungs or the brain. This could be fatal.

It's not known if blowing air into the vagina is also a threat to women who are not pregnant. But since many women may be in the early weeks of pregnancy without knowing it, the practice should be avoided.

You needn't worry about the small amounts of air that may get into the vagina during intercourse. That's very different from having a lungful of air forcefully blown into the vagina.

Sometimes air gets into the vagina if the woman assumes intercourse positions in which her pelvis is elevated—such as on her back with her knees close to her chest. Penile thrusting can also introduce some air into the vagina. Flatulentlike sounds may be heard as small amounts of air are expelled.

I had a vasectomy three years ago. Since then, I've been divorced and remarried, to a woman who has no children. Is there any chance that my vasectomy can be reversed so I can have more children?

Experts say there is some chance that surgery could restore your fertility. The testes usually continue to produce some sperm up to three years after a vasectomy. Even when a vasectomy is reversed within three years, however, fertility is restored only about 66 percent of the time.

A young man was found dead with a rope around his neck. It wasn't suicide. Rather, he had accidentally strangled himself while masturbating.

Apparently, he'd intended to shut off oxygen to enhance his orgasm. How common is this practice?

When the amount of oxygen to the brain is reduced by compressing the neck arteries, the resulting muscle spasm and distortion of time can seem to intensify an orgasm.

Cutting off the oxygen supply in the service of auto-eroticism is a very dangerous practice, however. If the victim becomes unconscious, or can't reverse the asphyxia, death by strangulation can result. Death is especially likely if the person applied pressure to his neck by assuming a ''hanging'' position.

Each year, ''sexual asphyxia'' causes about 250 to 500 deaths, estimates Dr. Myron M. Faber, on the basis of research he conducted with a colleague. In most cases, the victims are young men who have no history of psychiatric illness.

Dr. Faber, a specialist in adolescent medicine at Michigan State University College of Human Medicine at East Lansing, believes that these reported deaths represent only a fraction of the males who engage in this activity.

I often have fantasies about joining in group sex. I have these fantasies mostly when I'm making love to my wife or when I masturbate.

Lately, I've been going to adult bookstores and checking into group sex and even occasionally writing down the names of active couples. I want to pursue this but my wife does not. She sees group sex as a threat to our marriage.

Should I do it behind her back or just go crazy for the rest of my life? We've been married for ten years and I haven't strayed yet.

According to Dr. Harold I. Lief, a psychiatrist, your problem is not an uncommon one.

However, he says that if you cannot persuade your wife to join in, you will have to decide whether your marriage is more important to you than your wish to engage in group sex.

If it is, you will simply have to inhibit that activity. If you participate in group sex behind your wife's back, you will inevitably be found out, and your marriage may be destroyed. You should know, Dr. Lief adds, that for many people, fantasies about group sex are more rewarding than the actual experience.

How many abortions can a woman safely have?

Sex educator Judy Henkel reports that a properly performed pregnancy termination is one of the safest medical procedures. Most complications are minor. The risk of death from an abortion in the first trimester of pregnancy is less than one-tenth the risk of death in childbirth. The mortality rate for a suction abortion in the first eight weeks of pregnancy is 0.6 per 100,000 women.

While most physicians agree that a single, early termination does not threaten a woman's ability to bear children in the future, there is evidence that repeated abortions may pose such a risk. According to a study by the World Health Organization, women who have had two or more abortions are two to three times more likely than other women to have premature deliveries or low birth-weight babies.

The emotional aspects of repeated abortions are also an important consideration. Terminating a pregnancy is a serious matter, and it should not be used routinely as a method of birth control.

Occasionally I suffer from impotence that's caused by performance anxiety. Instead of relaxing, I focus on the problem and it gets worse.

Even though my partner is understanding, I don't know what steps to take to achieve an erection. Can you help?

Armando DeMoya, M.D. and Dorothy DeMoya, R.N., M.S.N., sex therapists, suggest that you focus on your partner's pleasure instead of your own when you find yourself impotent. The trick is to stop thinking about yourself. Instead, tell yourself, "If I get an erection, I get one. If not, at least making love will be enjoyable for her." If you pleasure your partner, you may well benefit from a "give-to-get" reaction: As her pleasure increases, you may become sexually aroused as well.

The best way to get over performance anxiety is to stop "spectatoring"—the sex therapist's term for watching yourself perform sexually and distancing yourself from your sexual feelings. It's best to try losing yourself in the moment instead of concentrating on how you're doing.

It may help you relax to know that the kind of intermittent impotence you describe is very common. It's often due to drinking too much alcohol, fatigue, and particularly to stress. The sex act should provide an opportunity to relax and get away from stress. So train yourself to think about your partner, or to think about nothing at all, and you'll be well on the way to solving the problem.

When I do real hard exercise, I sometimes ejaculate. Is there something wrong with me? (I'm 25.)

It's unlikely that you're ejaculating semen during your exercising. According to Mauro P. Gangai, M.D., associate professor of urology at the University of Texas Health Science Center, "True ejaculation, although probably not pathologic, would be very unusual under these circumstances."

Dr. Gangai thinks it's far more likely that the liquid coming from your penis is either a small amount of urine or a secretion from your prostate gland. "Either of these would not be very unusual in any age group during strenuous exercise."

I'm a 17-year-old girl. When I'm sexually aroused, I make a lot of noise. I don't mean to, but I gasp and pant. This annoys the guys I make out with. One of them even slapped me.
Am I normal, or am I a nymphomaniac?

When they're sexually excited, both males and females experience a faster heart rate. Blood pressure rises. And there is an increase in the rate of breathing. Thus, it's perfectly natural to breathe more quickly and shallowly with sexual arousal.

The gasping and panting you describe are particularly common in the plateau phase of the sexual response cycle, just before orgasm. At this time, heart rate, respiratory rate, and blood pressure rise even more.

The condition you describe is normal, and it is not nymphomania. Nymphomania is a medical term that describes, in women, a pathologically excessive need for sexual activity. In men, the condition is called satyriasis.

Women who have nymphomania constantly seek sexual gratification. Some of them are unable to reach orgasm. They look for orgasmic release with one partner after another. Others with the disorder may have multiple orgasms, but, nonetheless, never become satisfied.

Most commonly, hypersexuality is psychogenic in origin. The compulsive need for sex often compensates for feelings of inferiority. Nymphomania is considered a type of compulsive neurosis. In rare cases, however, brain disorders may account for the condition.

We hope you're reassured about your normalcy. But do give some thought to your choice of sexual partners. A boy who slaps you when you express physical responsiveness is neither a good lover nor a good friend.

Ever since I was 12 years old, I've been masturbating by using diapers. I enjoy it more when I urinate in the diaper and then masturbate while still wearing it.

I don't feel like a baby, and I've been happily married for seven years. My wife and I have "normal" sex, but I need more sexual release than she does. So I turn to the diaper.

I don't think there's anything wrong with the practice, but I wonder how widespread it is. Is there a name for it? How can I contact others who have this interest?

There are indeed other people who share your interest in diapers (and often other baby paraphernalia as well). The practice is called infantilism.

The Diaper Pail Fraternity (DPF)—Suite 164, 3020 Bridgeway, Sausalito, CA 94965—an organization of people with this special interest, claims a membership of more than 1,500.

The organization puts out newsletters, membership lists, reviews of diaper products, and narratives of personal experience.

I heard a friend talking about a "ping-pong" disease. What did she mean?

A "ping-pong" disease is an infection that's repeatedly passed back and forth between sex partners—like a ping-pong ball.

In a typical case, one partner, most often the woman, is treated for a sexually transmitted disease, a urinary tract infection, or a vaginal infection. Her partner, however, harbors the organism without showing any symptoms, so he doesn't get treated. After she is cured, and sexual relations resume, he reinfects her.

To keep this from happening, regular sex partners of people with genitourinary infections should be tested to see if they have the infection too. If they do, both partners should be treated simultaneously.

My wife is nursing our baby. Sometimes, when we make love, I like to suck on her breasts and drink some milk.

Am I taking away milk from the baby's supply?

On the contrary. Your sucking stimulates the breasts to produce more milk.

Breast milk production operates on the principle of demand and supply. As the growing baby sucks harder on the breast to get more milk, the pituitary gland responds by increasing the milk supply. In the same way, your sucking signals to the pituitary that more milk is needed.

How often should a boy my age (14) masturbate? I usually do it when I wake up in the morning, when I come home from school, and before I go to sleep at night. Sometimes I even do it four times in a day.

Do you think I masturbate too much?

For a boy your age, your sexual activity is "right on schedule," says Ralph I. Lopez, M.D., an associate professor of clinical pediatrics at New York Hospital–Cornell Medical Center in New York City.

"I don't know of anyone who has been able to quantify what the normal amount of adolescent masturbation is," says Dr. Lopez. "Certainly, during the rapid growth phase with significant sexual changes, I really couldn't get too concerned about three times a day or even four times a day."

Dr. Lopez reassures you that masturbation in and of itself is not thought to cause any harm. "It serves as an outlet for a sexual drive that is present in various forms from childhood to adult life."

So if you masturbate a lot, how can you tell if you do so excessively? More important than frequency is the part it plays in your life. Is masturbation just one aspect of your daily routine? Or does it serve as the entire focus of the day?

Dr. Lopez points out that masturbation is no problem for a teenager who is doing well in school, has adequate peer relationships, and is weathering the usual intergenerational conflicts with his family. But if there are difficulties in any of these areas, frequent masturbation may be a sign of underlying depression and anxiety.

What can you tell me about the screening test for AIDS (Acquired Immune Deficiency Syndrome)? If it's positive, does it mean that the person has AIDS?

The current screening tests for AIDS detect an antibody to human T-lymphotropic virus type III (HTLV-III), the virus that causes AIDS. People with positive tests should not donate blood. A positive test for antibody to HTLV-III, however, does not necessarily mean that the person has AIDS or will ever develop AIDS.

A positive test with no symptoms might indeed be due to the presence of AIDS. But it might also be caused by a subclinical infection, immunity, or a reaction with other viral antigens.

It might also be due to an error in the laboratory. So if a test is positive, it should be repeated at least one more time to help rule out error.

The test, at present, merely shows the potential presence of AIDS. In studies among homosexual men who had no symptoms but showed antibody to HTLV-III, about 50 percent remained asymptomatic after two to five years.

During the same time period, 5 to 19 percent developed full-blown AIDs. The remainder developed some conditions suggestive of AIDS, but these may or may not turn into AIDS.

If you take an AIDS test and it shows the presence of HTLV-III antibody, see a doctor immediately. Since many of the complications of AIDS are treatable, an early diagnosis may decrease the severity of the illness.

Be alert for early symptoms—such as persistent fever, diarrhea, headache, skin ulcers, shortness of breath, persistent cough, unexplained weight loss, or swollen glands.

Since many people with positive test results have the HTLV-III virus in their blood, semen, or saliva, they may be able to transmit the infection to others, even though they have no symptoms. They must be considered potentially infectious for an indefinite period of time.

If your test results are positive, you should not share razors, toothbrushes, needles, or other utensils that could be contaminated with blood. Any surface contaminated with your blood should be cleaned with household bleach, diluted in water. Inform your dentist, physician, and other medical personnel, so that they can take measures to protect themselves when treating you.

I heard about a kind of contraceptive that can give you protection for five years. Can you tell me more about it?

You heard about a hormonal contraceptive that is designed to provide up to five years of slow-release protection. The hormone is levonorgestrel, a progestin.

Six one-inch plastic capsules containing the hormone are implanted under the skin in the side of the woman's upper arm. Each day, the capsules release 30 mcg of the hormone, about the same amount of progestin delivered by the low-dose Pill or Mini-pill.

The contraceptive has been tested on 15,000 women. In the first nine months of what will be a five-year trial, it has been found to be more than 99 percent effective—more effective than the Mini-pill.

The five-year contraceptive is expected to gain FDA (Food and Drug Administration) approval by 1988. It will probably be marketed under the brand name Norplant.

The contraceptive works by preventing ovulation. About 60 percent of the women using it reported such side effects as irregular bleeding, longer periods, and longer intervals between periods.

Another 10 percent reported headaches, nausea, loss of appetite, dizziness, changes in sexual desire, weight changes, acne, and depression.

These side effects are typical of progestin.

My mother is 52, and she complains a lot about hot flashes. Are they real, or all in her mind? What can be done to help her?

The hot flash (or flush) is real enough. It is the most common symptom of the menopause.

The flashes are a feeling of heat spreading over the body. Visible blushing may start at the chest and continue up to the neck, face, and head. Then, it may spread all over. Although a woman can alternately sweat and shiver, if she takes her temperature, she finds that she has no fever. The flashes can show up as often as twenty times a day, and many women are awakened by them at night, too.

A woman may start to get hot flashes up to five years before her actual menopause (the cessation of menstruation). They can continue for several years thereafter.

Hot flashes are caused by the sudden excessive dilation of small blood vessels close to the skin's surface. This results in more blood being brought to the area, which produces local heat and activates sweat glands.

The instability of the blood vessels is believed to be due to changes in the production of various female hormones. The changes are spurred by the decrease in estrogen during menopause.

Your mother should discuss this problem with her doctor. Estrogen replacement therapy is often helpful in controlling hot flashes.

For a few minutes after having a climax, I find it unpleasant to deep kiss or to have my penis touched. It's as if my sexual energy has to recharge.

My wife is somewhat distressed by my reaction. Am I normal? If not, what could be the cause? I'm 37 years old.

You express the situation very well.

All men undergo a refractory period after ejaculation and orgasm that may last for a few minutes or many hours. During this period of "recharging," further ejaculation is impossible, although a partial or full erection may sometimes be maintained.

Just after ejaculation, men vary greatly in their response to further sexual touching. Like you, many men find it unpleasant to be kissed or fondled. A caress that was sublime just before orgasm can become intensely irritating afterwards.

Let your wife know that this is a common reaction. To satisfy her need for further intimacy after intercourse, experiment to find the kinds of touching that feel pleasant to you.

I'm considering becoming a surrogate mother. I am 35, white, healthy, and of northern European background. What are the going rates?

According to attorney Noel P. Keane of Dearborn, Michigan, the usual rate for surrogate mothers is about $25,000 plus medical expenses. In some locales, however, couples seeking a surrogate have advertised payments as high as $125,000.

A surrogate mother is most often sought by couples when the wife is infertile. The surrogate is artificially inseminated with the husband's sperm, carries the child to term, and gives it over to the couple after birth.

Before you decide to be a surrogate mother, explore the matter carefully. Serious questions about surrogate birth remain unresolved, many of them legal.

You may also be unprepared for the emotional realities of surrogate childbirth. You seem to be interested in the process for the money. But you don't know how you may feel if you have to relinquish a child—a child that is half yours—after carrying it for nine months.

To better prepare yourself, it may be a good idea to read Noel Keane's book, *The Surrogate Mother* (Dodd, Mead, 1981). And before you agree to become a surrogate, make sure you have your own lawyer to represent your interests.

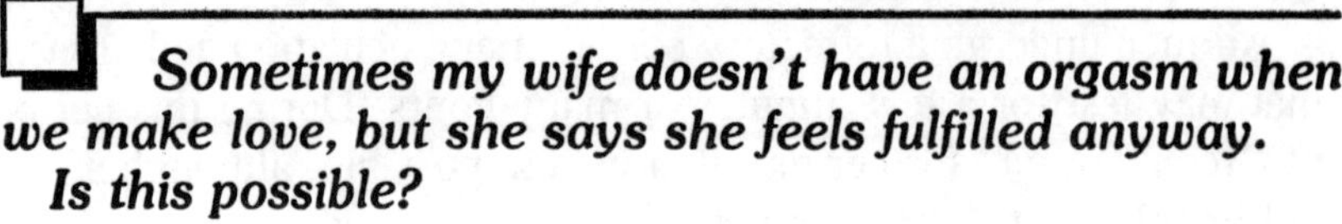

Sometimes my wife doesn't have an orgasm when we make love, but she says she feels fulfilled anyway. Is this possible?

Some women (and some men as well) find that they can on occasion have a perfectly satisfactory sexual encounter without feeling the need to have an orgasm.

"It is important to realize that many women, including those who are usually orgasmic during coitus, often comment on feelings of great pleasure from prolonged body contact," says Dr. Dorothy Strauss, clinical associate professor of psychiatry at the State University of New York, Downstate Medical Center.

The important thing, says Dr. Strauss, is how a particular woman feels about her relationship, her partner, and herself in such a situation. "As long as the woman indicates satisfaction in these areas, there is little reason to turn this matter into one of concern."

How soon after heart bypass surgery can I safely have intercourse? I'm a 51-year-old man.

It's important to discuss this matter with your doctor, since each case is different. Resuming sexual activity depends on how well a heart patient has recovered. Only a physician familiar with a particular patient's condition can give him specific advice.

Intercourse requires much less effort than is popularly believed. Often, heart rate rises more when driving a car in traffic than during intercourse.

In general, following heart surgery or a heart attack, a patient should resume less taxing forms of sexual activity on returning home from the hospital—touching, cuddling, and stroking. This gentle activity, without expectation of intercourse, helps relieve anxieties that can result in impotence. It can boost confidence and ease the resumption of intercourse.

A patient can safely go on to masturbation or oral sex as soon as he can tolerate an accelerated heart rate of 130 beats per minute. A heart patient can usually resume intercourse when he's able to climb one or two flights of stairs or walk several blocks at a brisk pace.

Your physician can assess how much sexual activity you can tolerate by giving you a standard series of stress tests followed by electrocardiograms. From the ECGs, the doctor can tell what your maximum heart rate can safely be, and he or she can estimate if your heart rate during intercourse falls within safe limits.

When you first resume intercourse after heart surgery, it's advisable to take it easy. The best time for intercourse is in the morning after a restful night's sleep.

Three positions which lessen cardiac workload are recommended: (1) lying on the side, in face-to-face position; (2) lying on the back, with the partner on top; and (3) sitting on a wide chair, low enough for the feet to touch the ground.

Refrain from having intercourse for several hours after eating a heavy meal or drinking alcohol.

Inform your doctor of any chest pain that occurs during intercourse. Also tell him about palpitations that last for a few minutes after intercourse, sleeplessness caused by sexual exertion, or marked fatigue on the day after intercourse.

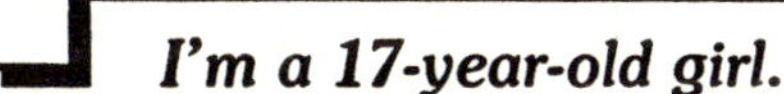

I'm a 17-year-old girl.
My doctor told me I have molluscum contagiosum, which he said was a venereal disease. But I've never had sex. So what's the explanation?

Molluscum contagiosum is spread by close physical contact, not necessarily sexual. It's a poxlike disease, caused by a virus.

Most cases occur among children and young adults. Epidemics sometimes break out in places where young people live closely together, such as summer camps or college dorms.

Molluscum is marked by flesh-colored or pearly papules with a depression in the middle of each. They generally appear on the face, eyelids, breasts, genitals, and inner surface of the thigh. Sometimes they occur on the mucous membranes of the mouth or rectum.

A doctor can remove the lesions by using a sharp curette, a scraping device. Bleeding is easily controlled with direct pressure, and anesthesia is not required.

If left alone, however, the lesions usually vanish spontaneously within a year.

My boyfriend has a garter and shoe fetish. He likes me to wear black garters and black six-inch-heel pumps to arouse him.

Our love play usually starts with my prancing around the room in my garters and heels. The tapping sound made by the metal heels on my pumps is also arousing to him. He finally gets very excited and makes love to me while I have the garters and shoes on.

I love my boyfriend very much, and want to please him. I don't mind catering to his fetishes. In fact, I feel sexy when I'm dressed this way. But I'm worried that I'm contributing to a mental illness he might have.

"Contrary to popular belief, a fetish doesn't necessarily indicate psychopathology," say Armando DeMoya, M.D., and Dorothy DeMoya, R.N., M.S.N. "As long as no one is truly suffering, there's little reason to alter the behavior."

A person with a fetish has an attachment to an object, or sometimes to a part of the body, that is charged with special erotic interest. The object (such as gloves, hair, or—as in your case—garters and shoes) becomes a primary focus of sexual satisfaction. As you describe, the object or objects are often incorporated into sexual activity with another person as an aid in sexual arousal. Sometimes the objects are used in masturbatory activity.

Fetishism is not generally considered to require corrective therapy unless it causes the person concern. From what you describe, your boyfriend does not seem concerned, nor are you reluctant to participate in the activity. In fact, you say it increases your own sexual arousal. Since the sexual scenario your boyfriend prefers pleases you—and hurts no one—there is no reason to discontinue it.

Nor is there any reason to be concerned about its "normalcy." There is very little agreement among sex researchers as to what constitutes "normal" sexuality. Say sex researchers William Masters and Virginia Johnson: "The variety of forms of expression of human eroticism is great. Defining what is sexually normal is highly problematic.

"Therefore, in discussions that deal with the boundaries differentiating normal from abnormal, it is best to be circumspect and to acknowledge openly the state of our ignorance."

My wife seems to develop a vaginal discharge shortly (within a day or so) after we have sex.

Her doctor has been treating her for a yeast infection, but the treatment doesn't seem to help. Could the cause be in my ejaculate? Is there such a thing as being allergic to semen?

Yes, there is such a thing as an allergy to semen. A small number of women experience allergic reactions to semen, with stinging, burning, and pain in the vagina. The vagina and vulva become red and swollen, and there are often hivelike bumps on the vulva.

In a few cases, the allergic reaction may be more extreme, with systemic symptoms such as swelling around the eyes, a sensation of swelling in the throat, or an asthma attack.

But it doesn't sound as if your wife has this problem. If she did, the allergic symptoms would start during intercourse or immediately after ejaculation.

It may be, however, that your wife is suffering from a "ping-pong" infection that passes back and forth between a couple by means of intercourse. Doctors recommend simultaneous treatment of a couple to break such a cycle.

It's also possible that your wife is suffering from more than one vaginal infection. This may explain why treatment for yeast infection alone does not resolve the problem.

My fiancé's penis is very large. This often makes intercourse uncomfortable for me. Are there any positions that would be better than others?

My doctor says I have no medical problems that could account for sex being painful.

The pain that women feel in such cases usually results from the penis pressing against the cervix or uterus. These organs are at the upper end of the vagina and can be sensitive to pressure.

It may be helpful to try intercourse positions in which the female partner can help control the degree of penetration. In the male superior (man-on-top) position the man controls the extent of penetration. In this position, thrusting with a long penis may cause discomfort.

Positions in which the woman has some control include the female superior position and the side-by-side position. In these positions, the woman can choose to receive only as much of the penis as is comfortable.

You should also make sure that you're sufficiently stimulated before penetration so that you become adequately lubricated. A well-lubricated vagina will help minimize discomfort from the girth of a very large penis. If you require more lubrication, use a water-soluble jelly, such as K-Y Jelly, or a contraceptive cream.

Is it safe to have sex with a person who has herpes, but is not currently having symptoms? Can condoms offer protection?

According to Nicholas J. Fiumara, M.D., a specialist in sexually transmitted diseases, it's possible to spread herpes without showing symptoms. Some people with herpes have asymptomatic reactivation of the herpes virus.

In fact, it's possible to have herpes—and to spread the disease—without ever having symptoms. At least 75 percent of patients infected for the first time have no symptoms at all. Yet they may nevertheless pass on the infection.

Using condoms properly and consistently is believed to provide some protection against herpes.

My 17-year-old daughter hasn't menstruated in several months. Her physician says that her periods have probably stopped because she's been practicing intensively for track and has lost a lot of weight.

Will this condition be permanent and interfere with her having children in the future? Also, does she need to use contraception, even though she doesn't have periods?

For a woman to menstruate, at least 10 percent of her body weight must be fat. Otherwise, her estrogen level will be too low for ovulation. That's why cessation of menstruation (amenorrhea) is often found among young women suffering from anorexia nervosa.

As with your daughter, amenorrhea is also common among otherwise normal women who are extremely lean, such as track team members, gymnasts, and ballet dancers.

In almost all cases, the amenorrhea is reversible. Soon after gaining the proper amount of weight, a woman will begin to menstruate. However, it may take some months for her periods to regularize.

If she has intercourse without menstruation, she still requires contraception, since ovulation can occur unpredictably.

Do you know how many people engage in oral sex?

The practice of oral sex has apparently increased considerably over the past few decades. "The most striking change in sexual patterns in recent years has been the increased acceptance of oral-genital sex," says Alfred Auerback, M.D., a psychiatrist.

Dr. Auerback, who is clinical professor of psychiatry at the University of California School of Medicine, observes that today "men and women participate in oral sex as a means of giving and receiving pleasure. Only for a few people does oral sex have negative or forbidden connotations."

Attitudes toward oral sex have changed considerably since researcher Alfred Kinsey did his pioneering study of American sexual habits in 1948. At that time, 45 percent of males said they had performed cunnilingus—oral stimulation of the female's genitals. In contrast, a recent study shows that as many as 75 percent of males under 35 with thirteen years of formal education say they practice cunnilingus. Of males over 35 with the same education, 66 percent report engaging in cunnilingus.

Kinsey found that 60 percent of married couples had participated in oral sex. Recent studies suggest that among young college-educated couples today, as many as 90 percent may practice oral sex. Research conducted by psychiatrist John P. Callan, director of the psychiatric clinic at St. Francis Hospital in Hartford, Connecticut, showed that fully 91 percent of people under 35 "considered oral-genital activity to be a part of normal sex."

A survey by *Redbook* magazine asked wives: "Have you ever performed oral-genital sex (fellatio) on your husband?" Only 9 percent of the respondents said "Never." Forty percent of them often performed fellatio, while 45 percent occasionally did so.

Such statistics have led Dr. Robert Athanasiou, a member of the Sexual Function Group at Albany Medical Center in New York, to state that oral-genital sex "is almost as common as the genital-genital variety."

I'm a 17-year-old guy.

Recently, my aunt came to visit us from San Francisco (she's 20). One night when my parents were out, my aunt got high on cocaine. She climbed into our hot tub— nude—and beckoned to me to join her. She pulled down my trunks and before I knew it, we were having intercourse. (I'm not saying it was right, but I'm no saint.)

The day after, she told me that if our relationship didn't continue, she would tell my parents that I had raped her. What should I do? Until I hear from you, I'll have to go along with her.

There's no doubt that you're in a tough situation. But your best bet is to tell your aunt that the sexual relationship between you has got to stop immediately.

Of course, you're ambivalent about that. You found the sex enjoyable, and you'd prefer to think that you're stuck with this situation. On the other hand, you know that this relationship has no future. Such a relationship—even if there are no blood ties between you and even if the state does not consider the matter incestuous—would be bound to have very unpleasant ramifications within your family if it became known.

To call your aunt's bluff, tell her that unless she agrees to keep mum, you'll tell your parents about the relationship first. Point out that your parents are more likely to believe you than her.

The power of that threat depends on the kind of relationship you have with your parents. Have you given them reason to trust and believe you in the past? Or have you done some things that would make them more likely to believe your aunt?

If you have any doubts that your parents would trust your word against hers, write an account of events as they actually happened. Don't go into explicit sexual detail. Instead, concentrate on your feelings of being trapped and your worry that your parents might believe her story. There's something very believable about events and feelings that are set down in writing.

You probably won't have to show your parents the letter, though. Tell your aunt that you've written it and that you'll use it if she insists on pulling the rape story. She's unlikely to want to create a family scandal. You'll probably find that she comes up with a reason to go back to San Francisco very soon.

My first baby was born by cesarean section. Can I have a natural (vaginal) delivery for my next baby?

"Once a cesarean, always a cesarean" is no longer true. The chances of your having a vaginal delivery are very good. Many women who have had cesareans have subsequently delivered vaginally.

Most often, the condition that made a cesarean necessary in one birth won't exist in the next. Talk to your obstetrician about the reason for the first cesarean, and ask about your chances for vaginal delivery next time.

In the past, vaginal births after cesareans were discouraged because of fear that the surgical incision would rupture during labor. Actually, this happened very rarely. Usually it occurred when the incision, made vertically in the upper half of the abdomen, had not healed properly.

The modern surgical technique is to make a horizontal incision low on the abdomen, very close to the pubic hair line. This shorter horizontal incision is much less likely to rupture.

The American College of Obstetricians and Gynecologists says it's alright to attempt a vaginal delivery after a cesarean if these criteria are met: The mother has the shorter, horizontal incision; the baby is in the head-down position; the baby weighs no more than eight pounds, eight ounces; the pregnancy has been completely normal; and the mother's condition is being monitored.

Even women who do not meet all these criteria might discuss with their obstetricians the advisability of vaginal delivery.

I'm a 44-year-old man, and I'm interested in the effects of aging on sex. What can I expect as I get older?

Recent research shows that, contrary to popular belief, many older people maintain active sex lives—sometimes into their eighties and nineties. In one study involving the elderly, an overwhelming majority of respondents said that they continued to be sexually active.

Most people experience gradual changes in their sexual responses as they get older. Sexual processes slow down with advancing age, often beginning in the forties, fifties, or sixties. In general, regular and frequent sexual activity helps to maintain sexual responsiveness.

Among the first changes a man might notice are slower and somewhat less firm erections. An older man is also less likely to experience spontaneous erections. He will probably need greater direct stimulation of his penis in order to attain erection.

One benefit of aging is that erections may last longer. An older man typically has an increased ability to maintain an erection for long periods before he ejaculates.

From a female partner's point of view, such sexual changes are often welcome. An older man's ability to maintain an erection gives her more time to come to orgasm during intercourse. Many women also find an older man's slower pace more satisfying, since he is more likely to engage in sexual play before insertion.

On the other hand, if an older man loses an erection before ejaculating, he may have more difficulty returning to a full erection.

An older man may also experience less need to ejaculate than formerly. Indeed, he may find that he can have satisfactory intercourse without ejaculating at all.

When he does ejaculate, he'll have decreased force and volume of ejaculate. There may be fewer expulsive contractions on ejaculating. There's also an increased refractory time in older men—a longer interval before he can have another erection and ejaculate again. In older men, the penis more rapidly returns to its nonerect state after ejaculation.

I'm a 25-year-old man with herpes type 2. I wear contact lenses and I've heard there's a problem with contact lens use and herpes.
What precautions should I take?

You've probably heard that a type of eye infection—herpes simplex keratitis—is caused by the herpes simplex virus. The virus attacks the cornea, causing inflammation, redness, scarring, and possible loss of vision.

You may be concerned about transmitting the genital herpes infection to your eye when you insert the contact lenses. Or perhaps you think that irritation by the lens could cause the virus to erupt in the eye.

Whatever your worry, Nicholas J. Fiumara, M.D., a venereologist, says you can be reassured. The use of daily contact lenses does not cause herpetic keratitis.

I've been using birth control pills for four years (I'm 26), and now I want to get pregnant.
Can I just stop taking the Pill and try to get pregnant right away? Or should I wait awhile before I try?

Many doctors see no need to delay conception after stopping the Pill.

''Birth control pills suppress ovulation and alter the endometrium only for the time in which they're being taken,'' comments obstetrician/gynecologist Philip D. Darney, M.D., of the University of California School of Medicine. ''Therefore, the next egg that an ovary releases has as much chance of being normal as any other, and the endometrium will be prepared for proper implantation if the egg is fertilized.''

Your own physician, however, may recommend that you use a barrier method of contraception until you have had one normal menstrual period after discontinuing the Pill. That's because it may take a few months for you to have a period after you stop using oral contraception. If you become pregnant before you menstruate, it's hard to date the conception.

Also bear in mind that, for reasons no one can explain, there's an increased incidence of twins in pregnancies that occur within the first three months after stopping oral contraception.

When I have oral sex with my girlfriend, her pubic hair catches in my braces. She doesn't want to shave. Is there anything that might help?

Randolph C. Myerson, D.M.D., an orthodontist and pediatric dentist in Kingston, New York, suggests that you ask your orthodontist about a product that football players wear to protect the mouth from braces in case of a blow. It's a thin plastic shell that fits over the teeth. The upper and lower parts are hinged in the back to permit free movement of the jaws and tongue. One brand on the market is Doublegard, by Masel Orthodontics.

Another alternative is dental gum. Again, ask your dentist.

I'm a 33-year-old man who has had diabetes for seventeen years. I often find that orgasm produces little or no ejaculate.

This seems to vary, depending on how well my blood sugar is controlled. I've heard of "retrograde ejaculation." Could you comment on it?

In retrograde ejaculation, semen is discharged backward into the bladder, rather than out through the urethra. Normally, the bladder neck closes during ejaculation, and contractions compress the urethra to propel the semen outward through the urethra.

But when the bladder neck fails to close as a result of muscle or nerve damage, the semen flows backward into the bladder.

Like you, a man may discover—after intercourse or masturbation—that while he experiences orgasm, there is no fluid ejaculate. The diagnosis is confirmed when sperm is found in the urine.

The most common cause of retrograde ejaculation is prostate surgery. Scarring may prevent complete closing of the bladder neck, allowing semen to enter.

As in your case, diabetes is another common cause of this condition. Spinal cord injury, pelvic fractures, surgery, or other trauma to the pelvis may also cause retrograde ejaculation. Some types of hypertension drugs can also be responsible.

Sexual performance is not affected by retrograde ejaculation in itself. But diabetes may cause sexual difficulties such as impotence. Talk over any such problems you may experience with your doctor. Various remedies may help.

Men with retrograde ejaculation need not be sterile, even though the condition is usually not correctable. If a man wants to have children, sperm can sometimes be collected from his urine and used to inseminate his wife.

I'm worried that I have incestuous longings for my eight-year-old daughter. I often get an erection when she sits on my lap. Occasionally, I also have sexual fantasies about her.

Lately I've been avoiding her. What should I do?

The erection you get when your daughter sits on your lap could be a reflex reaction. It's triggered by pressure on the surface of the penis, rather than by incestuous urges.

Armando DeMoya, M.D., and Dorothy DeMoya, R.N., M.S.N., sex therapists, advise you to make sure your daughter is seated so that she will not brush against your genital area. But that doesn't mean you should discontinue having her sit on your lap. Such displays of affection for a daughter (or son) are appropriate as long as the child welcomes them. Often, as children develop sexually, fathers push them away because they fear sexual arousal. That leaves the child with feelings of rejection and confusion (''What did I do wrong? Why is Daddy mad at me?'').

It's almost universal for fathers and mothers to have occasional fantasies about their children of either sex. But these fantasies are almost never publicly discussed because they make parents feel frightened and guilty.

Rest assured that any fantasy is okay. A fantasy can only be damaging to your child if you act on it. Since behavior controls are usually quite strong, that possibility is remote.

I'm a 22-year-old man. Since I started taking diet pills, I've been losing my erection during foreplay. Could there be a connection?

Indeed there could.

Most diet pills are amphetamines, prescribed to suppress appetite and promote weight loss. Although little research has been done on the effects of amphetamine use on sexuality, some men report episodes of impotence—the inability to achieve or sustain an erection. Decreased sexual desire has also been reported, particularly in women.

On the other hand, some men experience increased desire and enjoyment while using amphetamines.

If the problem persists, see your physician. You may have to alter your method of dieting.

If a woman were artificially inseminated with semen from a man who had AIDS, could she get it?

Scientists are trying to determine if artificial insemination could cause AIDS in a woman if the sperm donor carried the virus.

A research project funded by the University of California and San Francisco General Hospital will study lesbians who have been artificially inseminated. Since lesbians are not likely to have had much heterosexual contact, the presence of the virus antibody in them could be reliably linked to the donated sperm. Such research takes on new urgency in the wake of an Australian report that four women appear to have been exposed to AIDS through insemination from the same infected donor.

Researchers recommend that all potential donors of semen be given the blood test that detects an antibody to the AIDS virus.

I'd like to introduce sexual aids into lovemaking with my wife. Is a standard vibrator appropriate? Any suggestions or precautions?

Vibrators are usually made of plastic or rubber, and are either electric or battery-operated. Some are phallus-shaped. Others come with parts for various types of stimulation or massage.

Vibrators are sometimes applied to the genitals during masturbation, or as an alternative form of sexual stimulation with a partner. Self-stimulation of the clitoral area with a vibrator may be used as part of treatment for women who have never experienced orgasm.

Used with care, vibrators can give pleasure without harm. But there are some precautions you should take:

● Vibrators may be allergenic. Some people are allergic to a vibrator's rubber, plastic, or metal parts. You should be especially wary if you are subject to contact dermatitis from nickel jewelry or rubber gloves. Use of the device should be stopped at the first sign of genital reddening, swelling, itching, or blistering.

Warm-water baths or douches may relieve a vaginal inflammation. Persistent irritation requires a vaginal examination by a physician.

• Inserting a vibrator into the vagina or rectum may be hazardous, possibly causing bruising, tearing, or inflammation of delicate tissues. Before being inserted, a vibrator should be lubricated, then introduced slowly. While it is inside, the sex partner's weight should not press on it.

• To prevent infection, vibrators should be washed frequently with soap and hot water. If a vibrator is inserted into the rectum, it should be thoroughly washed before being applied to the genitals or inserted into the vagina.

• Frequent or prolonged use of vibrators can cause numbness of the genital area. This is usually temporary.

Is it safe for my partner to perform oral sex on me when I use a cream (or jelly) with my diaphragm? If it is safe, does it taste okay?

There's no need to worry about any danger to your partner from tasting contraceptive creams during cunnilingus—they're nontoxic. Jellies are safe, too.

As for taste: "We know of no brand that's especially pleasant to the tastebuds," say Armando DeMoya, M.D., and Dorothy DeMoya, R.N., M.S.N. Many of the DeMoyas' patients find contraceptive jellies—which are faintly minty—to be somewhat more palatable. "We're asked this question so often," comment the DeMoyas, "that we wish more contraceptive manufacturers would face up to the realities of oral sex and add a really effective flavoring to their spermicidal barriers."

At least one manufacturer has experimented with flavored spermicides (one was strawberry), but the taste was judged to be too medicinal.

I often find my mind wandering when I make love. Physically, I'm in bed with my wife, but mentally I'm holding business meetings.

I know I'm cheating both of us out of a better sexual experience. But I can't seem to keep the work problems

*out of my mind (I'm an executive with lots of job head-
aches). Is there anything I can do to turn off my head?*

Yours is a common problem, particularly among executives with
ongoing tensions.

Sex therapists usually suggest that people whose minds are preoccupied during lovemaking try to stay in the here-and-now by focusing on their partner's pleasure. Why not try concentrating on
providing your wife with pleasure and experiencing her reaction?
You might also leave a light on—and your eyes open—when making love. Otherwise, your mental screen is likely to tune in images
of your business day.

As soon as you detect your mind wandering off, regain your
concentration by whispering something to your wife—"I love you,"
for example. You're also likely to stay more attentive if you pe-
riodically alter your sexual activity.

If you're like many high-pressured executives, you may have
difficulty relaxing in general. You may feel guilty about "playing."
When you're not working, you think you're wasting your time.
These ideas can carry over into your sex life.

Sex therapists observe that a satisfying sex life requires the ability
to take time out. It's important to schedule time for unhurried
lovemaking. Take the phone off the hook, make arrangements for
the children, and spend a couple of hours without distractions.
Remember, your marital relationship is at least as important as your
work.

---

*Is it possible for a girl to get pregnant through cloth-
ing?*
*My girlfriend and I are in a coed college dormitory, and
do not believe in premarital intercourse. However, we do
have intercourse by rubbing our bodies together with our
clothes on. We wear our undergarments and our jeans.*
What are the chances of pregnancy in this situation?

Your girlfriend's chances of getting pregnant in the circumstances
you describe are very slight. Sperm are unlikely to penetrate the
layers of clothing. The thick jeans are a particularly good barrier.

However, it is important for you both to realize that it is possible for a woman to become pregnant without penetration having taken place. Sperm that is deposited on her vulva (external genitals) could make its way up her vagina and cause fertilization. Indeed, there are documented cases of pregnancy occurring in this manner.

It's also considered possible for a woman to become pregnant if her underpants are dampened by the ejaculate. Sperm are extraordinarily tiny, and they can penetrate openings in the weave of the fabric.

The sexual activity you describe seems to be mutually satisfying, and it avoids the hazards of pregnancy and venereal disease. But if you should ever engage in such activity without wearing jeans, take the extra precaution of ejaculating away from your girlfriend's genital area.

Is it possible for the penis to become damaged by rough handling, so that it's slightly misshapen when erect?

It's unusual for the penis to be damaged by sexual intercourse or masturbation. In rare cases, however, the chambers of the penis that fill with blood during an erection (the corpora cavernosa and the corpus spongiosum) may tear. Such a tear is often called a penile fracture, although no bones are involved.

The sudden application of external force to the erect penis can cause penile fracture. It can happen, for example, if a man rolls over on a full erection, or if the penis is forcefully kneaded in an attempt to reduce an erection. Adolescents have been known to experiment with bending an erection, thus causing a rupture.

Sometimes, the tearing occurs during intercourse as a result of a sudden shift in position. It may also happen when the female, astride the male, attempts insertion.

When the tearing occurs, there may be a crackling sound. The erection collapses, and the man feels intense pain. The penis appears distorted, and there may be discoloration. Sometimes, the urethra is injured as well.

Although ice packs can reduce swelling and pain, a penile fracture requires a visit to the doctor. Usually, anti-inflammatory medicine is prescribed along with a relaxant. Occasionally, a sharp angle on erection or other deformity persists and may require surgical correction.

If you feel that your penis is misshapen, see a physician. Conditions other than fracture may be responsible. One possibility is Peyronie's disease, in which scarlike tissue develops along the shaft of the penis and causes a bend in the erection.

I'd like to meet another lesbian, close to my own age (I'm 17), who wants a romantic, loving relationship.

How can I do it? I'm too young for a gay bar, and I wouldn't want to go to one anyway. It really seems unfair. It's so easy for heterosexuals to meet one another.

Dr. Fred Westendarp points out that young gays have unique problems. It's often difficult to meet gay peers, since many people are not willing to identify themselves as gay.

Gay teenagers don't know much about homosexuality in general. They don't know where to find older gays who would be positive role models—and, as a result, many gay youths spend years in ignorance, fear, and loneliness.

There are, however, encouraging signs of change. Gay youth groups now exist in many large metropolitan areas and offer support and social opportunities. Parents and Friends of Lesbians and Gays, a national support group, has chapters in most major cities. There are a multitude of church groups, representing every major denomination, that provide gay people with opportunities to worship and make social contacts.

On most university campuses, you'll find a chapter of the national student group, The Lesbian and Gay Academic Union. Several national periodicals provide gays with news coverage and classified ads as a means of meeting people. Gay hotlines in most cities list information about local activities.

The existence of these groups makes it easier for you to reach out to other gays. And reaching out to other people for love and support is a lifelong task faced by all people, whether they are gay or not.

It's common to let fear of rejection keep you from reaching out. But if you have a strong self-image, you'll be able to overcome that fear. Here are some suggestions for becoming proud of a key person—yourself:

• Keep the agreements you make—both with others and yourself.

• Be honest, as painful as this can be at times. It's important to speak the truth, if you want to maintain your integrity.

• Keep fit. Eat a proper diet, get plenty of exercise, don't smoke or use alcohol and drugs.

• Develop a positive self-image by reading about gays, socializing with self-respecting gay people, and seeking counseling, if necessary.

In his book, *Loving Someone Gay* (Signet, 1977), Dr. Don Clark says: "If you would find a lover (or friend), first make yourself into the kind of person you want your lover to be."

Once you're satisfied with yourself, you'll be better prepared to develop relationships with others. Here are some resources that gay and lesbian young people may find helpful in meeting people:

• Gay Youth Groups: The National Gay Task Force (NGTF) Hotline (800) 221-7044; Gay and Lesbian Youth of New York, 208 W. 13th Street, New York, NY 10014, (212) 834-0310.

• Parents' Groups: Federation of Parents and Friends of Lesbians and Gays, Box 24565, Los Angeles, CA 90024. New York: (914) 793-5198; San Francisco: (415) 347-7958; Houston: (713) 464-6663.

• Church Groups: United Fellowship of Metropolitan Community Churches (Gay Christians), 5300 Santa Monica Blvd., Los Angeles, CA 90029.

• Campus Groups: The Lesbian and Gay Academic Union, P.O. Box 82123, San Diego, CA 92138, (714) 295-4040.

• Periodicals: *The Advocate*, Suite 225, 1730 S. Amphlet, San Mateo, CA 94402; *New York Native*, 281 West Broadway, New York, NY 10013.

• Hotlines: Call local directory assistance and ask for Gay Hotline or Gay Assistance Line listings. For example, in New York City, there is the Lesbian Switchboard, (212) 741-2610.

● Resource Books: *Our Right To Love: A Lesbian Resource Book*, edited by Ginny Vida (Prentice-Hall, 1978; available from the NGTF and in most major libraries); *Loving Someone Gay*, by Don Clark (Signet, 1977); *Now That You Know: What Every Parent Should Know About Homosexuality*, by Betty Fairchild and Nancy Hayward (Harcourt Brace Jovanovich, 1979); *One Teenager in Ten: Writings by Gay and Lesbian Youth*, edited by Ann Heron (Alyson Publishing, 1983); *Young, Gay and Proud!* by the Gay Teachers and Students Group of Melbourne (Alyson Publishing, 1980); *The Lord is My Shepherd and He Knows I'm Gay*, by Troy Pery (Bantam Press, 1972).

CONFESSIONAL BOX
"I wish I could tell my mother"

Dirk, (26): I'm probably the happiest gay person around. Although some people feel compelled to tell their parents that they're gay, I don't. Some parents, because of their age, cannot accept their children's gayness. They grew up in another part of time. If I told my mother that I'm gay, I'd have to worry about her being unhappy for me, when, in fact, I'm very happy. She only has a few more years to live, and why should she spend those years adjusting to my life-style?

Nonetheless, I'd love to share my happiness with her. I met a guy through the computer. We've been together for fifteen months, and I'm proud of it. I want my mother to be part of that and realize that her son is happy, loving, and committed to one person. Parents have always wanted their children to choose doctors and lawyers as mates. Today, computer programmers are considered just as good. Ned, my lover, is a computer programmer and a fine one. Mother would be so proud!

I know that some boys have "wet dreams" (nocturnal emissions) and some don't. Do girls ever have orgasms in their sleep?

Yes, some girls and women do have spontaneous orgasms in their sleep. Sometimes the orgasms are associated with erotic dreams; at other times, they occur without a remembered dream. Like nocturnal emissions, a woman's sleep orgasms occur without direct genital stimulation.

As many as 70 percent of all the women in Alfred C. Kinsey's *Sexual Behavior in the Human Female* (Saunders, 1953) reported having dreams with sexual content. Less than half (37 percent) had dreams that culminated in orgasm.

Because a woman, unlike a man, reveals no physical evidence that an orgasm has occurred, some experts have questioned the accuracy of Kinsey's data. That skepticism is supported by the fact that only 3 percent of the women sampled in Shere Hite's *The Hite Report* (Dell, 1981) noted having dreams that ended in orgasm.

Some nights I really don't feel like having sex. I just want my wife to hold me and hug me before we go to sleep.
Is this normal? (We're both 26 and we've been married for almost a year.)

Yes, it's very normal and common for couples to want to hold and touch each other without going on to intercourse.

Unfortunately, many people—particularly men—have been influenced by the myth that touching must lead to sex. This notion, comments Dr. Bernie Zilbergeld, a clinical psychologist at the Human Sexuality Program of the University of California, "robs us of the joys of just touching."

Dr. Zilbergeld points out that the simple pleasures of touching —hugging, cuddling, kissing, holding, and caressing, without proceeding to intercourse—are rarely portrayed in the media. Such kinds of touching are "completely absent in pornography and rarely encountered in any erotic material." Instead, touching is always portrayed as the first step toward intercourse.

''Touching is not seen as something pleasurable in its own right,'' says Dr. Zilbergeld, but is thought useful only ''to the extent that it paves the way to a presumably grander event.''

It's no wonder that you're concerned about the normalcy of simply wanting to be held. Men are conditioned to the idea that touching must be either sexual or aggressive. According to what they've been taught, it really isn't OK for a man to want only a hug or kiss, although many men experience the desire you describe.

I say that love bites can be dangerous. My husband says that's a myth.
Who's right?

You are.

If the skin is broken, a human bite is even more infectious than a non-rabid animal bite. That's because such a bite can transmit mouth flora that are resistant to human antibodies.

A human bite that penetrates the skin is best treated as a potentially serious medical problem. Wash the wound thoroughly with soap and flush it with running water. Apply an antiseptic solution and an antibacterial ointment. Then put a bandage over the wound.

I'm a 19-year-old guy who's been dating a 17-year-old girl for the past two years. We've been having sex for the past year.
Recently, she has wanted to have sex in public. We have done it at the movies. But now she wants to do it on buses and the subway.
I don't want to go along. What should I do?

It's unwise for you to go along with her. First, because you don't want to. Second, because you're likely to be caught—and possibly arrested for indecent exposure and public lewdness. Another important consideration: Since the girl may be under the age of consent in your state, you also risk being prosecuted for statutory rape.

We don't know just what psychological need accounts for your girlfriend's wish to have sex in public places. She may feel sex is most exciting when there's an element of danger and risk involved. Or she may enjoy the exhibitionistic aspect of having intercourse in public.

Another possibility is an unconscious wish to be caught and punished for the act. It's possible, though, that your girlfriend simply has poor judgment.

Whatever her motivations, you can be firm in your refusal to have sex in public.

I had a vasectomy at age 31, and a reversal last year at age 37. I now have a low sperm count, and the sperm check indicated low motility as well.

My urologist says my sperm count is too low for my new wife to conceive. Is there anything that can be done?

When a man has a low sperm count, it may be possible to achieve a pregnancy by obtaining a split ejaculate. In this procedure, the first part of the ejaculated semen—which has the highest number of active healthy sperm—is caught in a container and transferred directly into the cervical canal.

Another alternative is test-tube fertilization. Your wife's egg would be withdrawn from her ovary, fertilized with your sperm in a test tube in the laboratory, and injected into her uterus. This technique is still experimental and is subject to a high rate of failure.

Your urologist may want to refer you to a fertility specialist or clinic for such procedures.

I'm embarrassed because I get sounds in my vagina during intercourse that sound like gas.
What causes them, and what can I do about them?

Vaginal sounds during intercourse are very common. Like you, many women are greatly embarrassed by them.

A likely cause for the vaginal sounds is that you're assuming intercourse positions in which your pelvic area is elevated. You may have your hips on a pillow, for example. Or you may be in a position in which your knees are close to your chest.

In such positions, a small amount of air may enter the vagina. Then, when you change your position, the air is forced out. Similar flatulentlike sounds are often heard in female exercise or yoga classes.

Penile thrusting may also introduce air into the vagina. The consistent motion also forces the air out, contributing to the sounds.

Vaginal sounds during intercourse are perfectly normal. There's no reason to try to avoid them by giving up positions you enjoy. Since an elevated pelvis tends to permit deep penile penetration, it's very gratifying to many couples. Instead of being embarrassed by vaginal noises, try to accept them as a usual part of lovemaking.

What are you supposed to do when you find your son masturbating?

That depends a great deal on the age of your child.

If you come upon your preschooler playing with his genitals, explain that you know it's enjoyable, that many people do it, and that it's okay with you. But you should tell him that, like other bodily functions, touching the genitals is something people do in private.

Even small children can accept such logic easily. It's similar to learning to keep their clothes on, even though being undressed feels more natural.

If you discover an older child masturbating, chances are you're not giving him enough privacy. Simply say, "Sorry," and close the door. Take care to give your children bedroom and bathroom privacy. Always remember to knock before opening a closed door.

How many teenage girls get pregnant? How many have abortions?

Each year, over a million teenage girls become pregnant—about one in ten teenage girls nationwide.

Three out of ten teenage girls who have intercourse become

pregnant. About 30,000 of the pregnant teens are under 15 years old.

About a third of pregnant teenagers end their pregnancies with abortion. The younger a girl is, the more likely she is to opt for this choice. Among pregnant girls 14 and under, there are more abortions than deliveries.

A teenage mother who does not end her pregnancy is likely to keep the child. Of the more than 200,000 out-of-wedlock babies born to teenage mothers each year, the overwhelming majority are not given up for adoption. According to the National Center for Health Statistics, about 94 percent of babies are kept by their teenage mothers.

My wife was recently treated for pelvic inflammatory disease caused by chlamydia.

I always thought chlamydia was sexually transmitted. Since I have been faithful, I wonder if my wife has had an affair. Is it possible to contract chlamydia without intercourse?

Give your wife the benefit of the doubt.

Most cases of chlamydia are transmitted through vaginal or anal sex with someone who has the infection. But the infection can also apparently be picked up in contaminated swimming pools. It may also be transmitted to the eye by a hand moistened with infected secretions.

Chlamydia infections are caused by *Chlamydia trachomatis* organisms, bacterialike parasites that multiply within human cells.

Chlamydia is responsible for more than half the cases of nongonococcal urethritis (NGU)—inflammation of the urethra caused by organisms other than the gonococcus bacteria. Symptoms may appear one to three weeks after contact. Since chlamydia symptoms are similar to symptoms of gonorrhea, the condition is often misdiagnosed.

Symptoms in men include a burning sensation on urinating, and a urethral discharge. But about 10 percent of men have no symptoms, even though they can transmit the disease. In women, symptoms may include painful urination, a thin vaginal discharge, and lower abdominal pain. Many women, too, show no symptoms.

Women with untreated chlamydia can develop urethral infection, inflammation of the cervix, pelvic inflammatory disease, and infertility, as well as complications during pregnancy and birth. It's a serious condition.

Frequently only one member of a couple will have symptoms, although the other harbors the infection as well (this could be true in your case). Both partners should be treated to avoid passing the infection back and forth in a "ping-pong" fashion.

Tetracycline is the standard treatment for chlamydia infections. Sulfa drugs are an alternative. The infection usually clears up within three weeks.

I've been cross-dressing for about twenty years (I'm 37 now).

Most of the time, I just wear women's silk lingerie under my suits. But sometimes, when my wife is away on business trips, I go out and spend hundreds of dollars on women's clothes.

I wear the stuff maybe twice before I panic, thinking I'll be found out. Then I get disgusted with myself, throw away all my beautiful, secret clothes, and vow never to do it again. But before too many months pass, I yield to the craving again.

Do other transvestites go through this sort of thing? Can anyone cure people like me?

Apparently, your pattern is common among cross-dressers.

The Gateway Gender Alliance—which offers information and support for transvestites and transsexuals—says that many cross-dressers go through the periodic purges you describe. "Their wardrobe is discarded and they swear never to cross-dress again! But within a few days, weeks, or months they are accumulating a feminine wardrobe and doing it again."

Many methods for curing transvestism have been attempted, but with little success. Some therapists try psychoanalysis; others use aversion therapy with shock treatment, behavior modification, tranquilizers, or desensitization. There is one method that has had some success, however. Sex therapists Dr. Wardell Pomeroy and Dr. Leah Schaefer give the transvestites they treat permission to cross-dress. They also counsel their patients on where and when it may be best to cross-dress.

The therapists' support and acceptance leads, at first, to an increase in the behavior. Over time, however, the compulsion to cross-dress diminishes and, in some cases, it even disappears.

Pomeroy and Schaefer act on the theory that transvestism is not harmful. They urge patients to accept the idea that their problem is social, not psychiatric. As soon as patients achieve this acceptance, their transvestism diminishes in importance and they obtain relief from guilt.

I always thought that men got morning erections because they need to urinate. However, I often find that I have an erection in the morning but hardly have to urinate at all.
What accounts for this?

It's only coincidental that a male who awakens with an erection also has to urinate.

In healthy males, erections occur several times during the night as part of the cycle of rapid eye movement (REM) sleep. A typical 25-year-old male will have about four erections a night. The final episode usually occurs just before normal wake-up time, so when the sleeper arises, he has a so-called morning erection.

Constance Moore, M.D., assistant professor in the department of psychiatry at Baylor College of Medicine in Houston, says that morning erections are not due to erotic dreams or unsatisfied sexual desire, as is commonly believed. "These beliefs have little validity," she comments. REM-related erections are affected little, if at all, by "anxious dreams, obvious neurosis, or the amount of sexual satiation before sleep."

There's some dispute among sex therapists as to what constitutes premature ejaculation.

Certainly, if a man ejaculates before his penis enters the woman's vagina, the ejaculation is considered to be premature. But what's premature after entry? Two minutes? Five minutes? Ten minutes?

In general, premature ejaculation is thought to be a chronic problem if a man cannot maintain control long enough to satisfy his partner at least half the time—assuming that she can attain orgasm through sustained intercourse.

But since most women do not routinely reach orgasm through intercourse, sex therapist Helen Kaplan may have a more useful definition: A premature ejaculator is a man who does not realize when he is almost ready to ejaculate. Therefore, he cannot delay his ejaculation. This definition emphasizes ejaculatory control rather than the amount of time involved or the woman's ability to reach orgasm.

Treatment for premature ejaculation focuses on making a man more aware of the sensations that precede orgasm. As he recognizes these sensations, he can learn to modify his responses in order to delay ejaculation.

The remedies you ask about—anesthetic ointments and distracting, boring thoughts—have not proved to be effective. Using an ointment does not help a man learn how to control his ejaculation. Furthermore, it can rub off on the woman's clitoral area, reducing her sensation and diminishing her pleasure. Trying to distract yourself with boring thoughts is also likely to be counterproductive as it does not produce ejaculatory control.

I'm a 27-year-old unmarried woman.

Although I use a diaphragm for contraception, I know I'd be much safer if my lovers used condoms. But men don't want to use a condom. They say it interferes with their sensation and interrupts lovemaking. How can I encourage a man to use a condom?

You're right in thinking that the combination of a diaphragm and condom is excellent protection against pregnancy. Like birth control pills, it's about 99 percent effective.

The condom can also help protect you against a variety of sexually transmitted diseases—particularly gonorrhea, which is widespread in the young, unmarried population.

Men may be afraid of losing their erections when they stop to put on a condom. This is less likely to happen if fitting a condom over the erect penis is turned into an exciting part of sex play. Try to devise ways of making the procedure stimulating. Remember that a lubricated prerolled condom with a reservoir tip is usually the easiest to put on. Another way to make condoms more sensuous is to suggest that your partners try buying them in a variety of designs and colors.

Males who seek greater sensitivity are often satisfied with condoms made from lambskin or extra-thin latex. A wide range of sensations may also be experienced by using condoms embossed with various textures.

I've read that it's sometimes possible for a woman to nurse her adopted baby. I expect to adopt a baby soon, and I'd love to learn how to nurse it.

Can you tell me more about this method?

A promising method of inducing lactation in adoptive mothers is provided by the Lact-Aid Service and Supplies Center, Box 1066, Athens, TN 37303. The company sells a nursing kit that comes with information on inducing breast milk supply. The Lact-Aid kit allows the baby to get an adequate milk supply while it suckles the breast, thereby stimulating the breast to produce milk.

The kit consists of a pre-sterilized disposable bag with infant milk formula and a nipple. The woman fastens the bag to a nursing bra or suspends it between her breasts. A length of soft, fine flexible tubing extends from the bag and slowly dispenses the formula. As the baby sucks from the formula bag, he can also be put to suck at the breast, providing it with the stimulation needed to induce lactation and produce a natural milk supply.

Not all women can produce breast milk in this way. Even if you are able to, don't expect to completely dispense with the formula and provide your adopted baby with its total milk supply. Most adoptive mothers must continue to supplement their breast milk supply.

Some breast milk usually appears within four weeks of beginning to nurse the baby with the Lact-Aid device. Peak breast milk production occurs about twelve weeks after the onset of lactation.

In one study, the best results were obtained when the babies were under a month old when they began to nurse. You're also likely to be more successful if you have the full support and encouragement of your husband, and if you educate yourselves about breastfeeding in general. A good source of information and support is La Leche League International, 9616 Minneapolis Avenue, Franklin Park, IL 60131, (312) 455-7730.

I'm a transsexual. I've been on hormones for three years, and now I'd like to have the sex-change surgery. But I've heard that it costs many thousands of dollars.

Do health insurance companies offer coverage for this operation? Are there any foundations that might help out, the way they do for transplants?

According to *Guidelines for Transsexuals* (a booklet distributed by the Janus Information Facility, 1952 Union Street, San Francisco, CA 94123), no private foundation offers financial assistance to transsexuals for surgery.

For those who qualify for veterans' benefits, Veterans' Administration Hospitals usually provide only postoperative care. Surgery may sometimes be financed through bank loans, if you have sufficient collateral.

Your best bet is a good health insurance policy. *Guidelines for Transsexuals* advises that you read the insurance contract carefully before signing. Some companies specifically exclude treatment for transsexualism from their coverage. Also, most policies stipulate a twelve-month waiting period before providing benefits for previously diagnosed conditions.

Be sure to apply for and sign the contract with your birth name and sex. If you fail to do so, coverage may be terminated on grounds of fraud.

Some health insurance companies state that the holder is covered only for "necessary treatment of an injury or disease process." In such a case, your physician should represent transsexualism as a distinct, medically definable disease entity for which treatment is required. It's advisable for you and your physician to examine the wording of your policy, so he or she knows how to frame the diagnosis.

Your doctor should not describe the treatment simply as "transsexual surgery" or "cosmetic surgery." Reimbursement for these categories is consistently denied. The best results are obtained by stating that transsexualism is a neuroendocrinological or psycho-hormonal disorder that requires surgical and hormonal treatment. Another effective way to classify the condition is as "gender dysphoria."

Recently, some insurance policies have become more liberal in providing coverage for preoperative evaluation, sex reassignment surgery, and related therapies.

I'm almost positive that my 15-year-old daughter and her boyfriend are having sex. If I tell her to get on the Pill, she'll think that I approve, which I definitely don't.

But I'm worried sick that she'll get pregnant. Could I get birth control pills and put them in her food?

Judy Henkel, a sex therapist certified by the American Society of Sex Educators, Counselors, and Therapists, says no. The birth control pill is a drug that must be prescribed for an individual by a physician. Because not all women can safely use the Pill, a medical history must be taken, laboratory tests performed, and a physical examination conducted.

You need to communicate your concerns to your daughter in a calm and nonjudgmental manner and help her to make a decision about her sexuality that will be in her best interests. By putting medication in her food, without her knowledge, you would jeopardize her health and violate her rights, without addressing the issue.

Many young people are involved in sexual activity at an early age because of low self-esteem and peer and media pressure. Many of them lack information about pregnancy and sexually transmitted diseases. By providing your daughter with information and discussing family values about sexuality, you can help her to better understand her sexual self. If you find that contraceptive counseling is a necessity, make an appointment with your physician or family planning clinic.

I'm a 30-year-old man and I can't seem to have an ejaculation no matter what I do.

I've tried everything—bondage, vibrators, all sorts of things. But nothing works. What should I do?

You may be suffering from a relatively uncommon sexual dysfunction called ejaculatory incompetence or inhibited ejaculation. It's the inability to ejaculate in the vagina.

Some men suffering from this problem may be able to ejaculate through masturbation, or with oral or manual stimulation, even though they are unable to ejaculate during intercourse. In a small number of cases, a man may be able to ejaculate with one woman, but not with another.

Typically, men suffering from retarded ejaculation have normal sexual desire and erections. They can usually maintain firm erections during intercourse for a long time.

There are two types of ejaculatory incompetence—primary and secondary. In primary ejaculatory incompetence, the more common form, a male has never ejaculated in intercourse. The causes, most sex therapists currently believe, are usually psychological. Such men often have had a strict, antisexual upbringing. They may harbor feelings of guilt and shame about sex.

In secondary ejaculatory incompetence, a man has been able to ejaculate in intercourse before developing the condition. In many cases, inhibited ejaculation is a symptom of problems in the relationship.

Sometimes, for example, a wife wants to become pregnant, but if a husband fears fatherhood, he may unconsciously withhold his ejaculate. Some men can trace the development of ejaculatory incompetence to a specific traumatic event, such as the discovery of a wife's infidelity. Hostility toward a sex partner may also contribute to the problem.

While most cases of ejaculatory incompetence are psychogenic, organic problems sometimes account for the difficulty. Neurologic disorders may interfere with sympathetic nerves to the genitals. Some medications may inhibit ejaculation, as can chronic alcoholism.

You need a thorough medical evaluation to determine the cause of your problem. If it's found to be of psychogenic origin, treatment by a sex therapist may help.

I am 17 and gay, and just started my first relationship. Both my partner and I enjoy orally stimulating each other's anuses with our tongues.

Is this dangerous? We keep very clean and always wash before doing it.

Fred Westendarp, M.D., says that there are definite dangers associated with anilingus or anilinction (stimulation of the anus by the partner's tongue, sometimes called "rimming"). You should be aware of the risks involved.

In a relationship where both partners are monogamous and neither partner carries a communicable disease, anilingus is probably not too risky. Even so, a number of intestinal infections—amebiasis, giardiasis, and shigellosis, for example—are known to be spread by anilingus.

The symptoms of these intestinal infections vary from person to person. They often include fever, diarrhea, abdominal cramps, bloating, gas, nausea, fatigue, and loss of weight. However, some people have mild symptoms or no symptoms at all. They feel and appear healthy and can pass the infections on to others.

Diarrheal illness is often difficult to diagnose, and often requires many visits to the doctor for expensive stool examinations. That's why it's important that your doctor be familiar with your life-style so that he can perform the proper tests. Treatment doesn't always kill the infecting organisms immediately, either. You must be tested again to make sure that you are cured and no longer infectious.

If you are not willing to give up anilingus altogether, intestinal infections can be limited by:

- Washing the anus carefully prior to anilingus
- Limiting the number of partners with whom you practice anilingus
- Being careful not to transfer secretions to your mouth after inserting the penis, a finger, or a dildo in the anus. Wash anything that has been in a rectum carefully before you put it in your mouth.
- Avoiding anilingus until any infection you have acquired is cured and your doctor gives you the go-ahead.

Hepatitis A (infectious hepatitis) is spread by the fecal-oral route, and practicing anilingus places you at high risk for this disease. Other infections that can be spread by anilingus—although this is not their most frequent method of transmission—include herpes simplex, syphilis, hepatitis B, and possibly AIDS.

CONFESSIONAL BOX

"Coming out has taken the pressure off"

Kevin, (22): I made sure that I was really committed to the gay life before I "dropped the bomb" on my parents. I never actually had to tell them anything; I just took my lover home and introduced him as a friend. I'm sure my parents know, but nothing is ever mentioned. My family and my lover get along great. Coming out has taken all of the pressure off my shoulders. It feels very good to wake up in the morning and have someone who you really love wake up next to you. I just wish the rest of those coming out could feel as I do. Good luck!

I am a 29-year-old woman, and I use an IUD (intrauterine device) for birth control.

I've heard that doing yoga exercises could dislodge an IUD. Should I avoid yoga?

Mary Anne Friederich, M.D., a gynecologist, says she is not aware that yoga, or any other exercise, can dislodge an IUD.

I have lumps in my semen (my wife calls them "tapioca"). What could be the cause of them? Is there anything to worry about? I'm 29 years old.

Terrence Malloy, M.D., a board-certified urologist, says that the consistency of the ejaculate (semen) may be affected by several factors. If a person has an infection in his prostate or seminal vesicles, the semen may seem thicker and at times be discolored to a reddish brown (hematospermia) or a deeper shade of yellow.

Also, some people lack certain enzymes, which makes it difficult for their semen to liquefy.

If you continue to have problems with very thick, lumpy semen, consult a urologist for a complete examination.

I've been having an affair with my first cousin.

Is it incest? Is it legal for us to marry? If we have children, will they be retarded or anything?

According to Sheldon Reed, author of *Counseling in Medical Genetics*, slightly over half the states prohibit marriage between first cousins. But most states do not legally consider intercourse between cousins to be incest.

Predicting abnormalities in the children of people who are related by blood is highly complex. Some conditions tend to increase your risk of having children with genetic abnormalities.

You're at greater risk, for example, if you and your cousin share several lines of common ancestry. That's probably the case if your ancestors came from a small, isolated community where marriages within the family were common. You're also at greater risk if you come from an ethnic group or family with a high frequency of genetic abnormalities.

Should you and your cousin consider marriage, get the advice of a professional genetic counselor. He or she can take your family histories and estimate your risks of having children with genetic abnormalities. To find a counselor, ask your family physician or get a recommendation from your local hospital. You can also ask a local chapter of the March of Dimes Birth Defects Foundation or contact the national headquarters at 1275 Mamaroneck Ave., White Plains, NY 10605; (914) 428-7100.

I'm a 66-year-old man facing surgery for an enlarged prostate. What effect will this operation have on my sex life? I've heard that it can cause impotence.

There should be no loss in sexual function following your surgery.

The operation is called a prostatectomy, which means removal of the prostate. Actually, the prostate itself is not removed, merely the mass of overgrown urethral glands inside the prostate.

You will most likely be able to have an erection, engage in intercourse, and experience orgasm after the surgery. But the operation does usually cause infertility because, after it, most men experience retrograde ejaculation: The semen is propelled into the bladder instead of being ejaculated through the penis. Since most men who have the surgery are older, infertility is not a common concern.

There is no physiological reason for a man to become impotent after a prostatectomy. But some men equate fertility with virility. If they feel less manly after the operation, they may have difficulty obtaining or maintaining erections. Or a man may fear that his sex partner considers him less virile since he can no longer father a child. Some men become disturbed because they cannot discharge semen upon orgasm. Any or all of these psychological factors may result in impotence.

What's the average size of penises? I'm worried that mine isn't big enough.

You're not alone in your concern. Yours is the question we're asked most frequently.

Alfred Kinsey's pioneering sexual research produced tabulations of the measurements of 2,500 erect penises of American men. Almost 90 percent were within an inch of the average six-inch erection.

In *McCary's Human Sexuality*, Dr. James Leslie McCary and Dr. Stephen P. McCary note: "In the average American male the penis is 2.5 to 4 inches long when flaccid (limp), slightly over 1 inch in diameter, and about 3.5 inches in circumference." When erect, "the average penis extends 5.5 to 6.5 inches in length, and becomes 1.5 inches in diameter and about 4.5 inches in circumference."

Dr. Michael Carrera, in *Sex: The Facts, The Acts and Your Feelings*, offers essentially the same figures. He says that an erect penis is usually between 5 and 7 inches long with a diameter of 1.25 to 1.5 inches.

Penis size is not related to body shape, height, length of fingers, length of nose, or anything else.

Researchers emphasize that there's great variation in the size of normal penises. "The measurement of a perfectly functioning erect penis can vary from 2 inches in one man to 10 inches in another, with one no less capable of coital performance than the other," say the McCarys. Once more, there's little relationship between the size of a flaccid penis and the size it will be when erect. A penis that is small in its flaccid state can show a remarkable change in size as it erects. By contrast, a large flaccid penis often changes very little in length or thickness as it becomes erect.

Why are men so concerned with the size of their penises?

"It does not take much imagination to understand why penis size often takes on great importance," comments Dr. Robert Crooks in *Our Sexuality*. "As a society, we tend to be overly impressed with size and quantity."

Men often assume that the bigger a penis, the better. They mistakenly equate greater penis size with greater masculinity. What's more, Crooks contends, if either a man or his partner view his penis as being too small, their sexual satisfaction will be decreased—not because of physical limitations, but as the result of a self-fulfilling prophecy.

How important is penis size to a woman's sexual pleasure? Some women do indeed express a preference for a large penis. But many more do not.

One woman told Dr. Crooks: "Penis size is important to my pleasure, but not in the way you might imagine. If a man is quite large, I worry that he might hurt me. Actually I prefer that he be average or even on the smaller side."

In other research, 75 percent of women surveyed by *Forum* magazine stated that penis size did not affect their relationships with their partners. "A man's character is far more important than his penis size," said one. Over half the women surveyed said that a medium-size penis was preferable to a large one.

So worrying about penis size can only hamper your pleasure— and size is probably of no concern to your partner.

I've heard that you can get herpes from toilet seats and hot tubs. Is this true?

Possibly.

Scientists used to think that herpes was transmitted only by sexual or other intimate contact because the virus that causes the disease could not survive for long outside the body. But new research shows that the herpesvirus can survive for minutes or even hours. That could mean herpes may be transmitted through toweling, toilet seats, and hot tubs. A recent report in the *Journal of the American Medical Association* presents evidence that such transmission is possible.

After several health clubs in the Washington, DC area were closed because of the suspicion that customers had been infected by herpes in hot tubs, investigators duplicated hot tub conditions in the laboratory. They found that "significant amounts of virus survived for 15 minutes" in water at 100 degrees Fahrenheit that contained the same amount of chlorine used in hot tubs. On the tub's plastic surfaces, the herpesvirus survived more than four hours.

The investigators concluded that hot tubs are "important possible routes for nonvenereal spread" of the herpesvirus.

Is it okay to smoke pot while I'm pregnant? I'm embarrassed to ask my doctor.

I'm 22, and this is my first pregnancy.

You're wise to be wary.

Research on the effects of marijuana use during pregnancy has been limited. But animal experiments clearly show that marijuana's principal ingredient, THC, crosses the placenta and enters the bloodstream of the fetus. Thus, marijuana has the potential to cause birth defects, and doctors caution against its use during pregnancy.

However, since there is no direct evidence that marijuana causes birth defects, therapeutic abortions for women who have used it during early pregnancy are not advised.

CONFESSIONAL BOX

"Stay off coke"

<u>Michael, (31)</u>: In response to the discussion that's been going on about cocaine and sex, I'm a former heavy user of about half an ounce of pure stuff every week. That's about one and a half ounces on the street. While I was on coke, sex was impossible. The desire was very strong—but my body was unable to comply. I'm clean now, and my sex life and my marriage are safe. I don't recommend even one or two lines of coke as a prelude to sex. Instead, try a glass of wine. It's a lot safer and it's not addictive. Doctors may say that coke isn't addictive, but I know better.

I know that blood pressure medications (antihypertensives) can sometimes cause impotence in men. Can it cause sexual problems in women?

I'm a 43-year-old woman, and I've just started to take medicine for high blood pressure. My doctor was evasive about the sexual issue.

Some women do experience sexual side effects from high blood pressure medications.

Taking antihypertensives may occasionally result in decreased vaginal lubrication and the inability to reach orgasm. Some women on blood pressure medication also experience a decrease in sexual desire. But the problems can often be relieved by modifying the dosage or switching from one type of antihypertensive to another.

Often the sexual difficulty does not arise from the drug alone. It may be compounded by fatigue, marital discord, alcohol consumption, and concurrent illnesses and medications. When these contributing causes are modified, a woman's sexual functioning can be restored even if she stays on the antihypertensive.

Is it possible to decrease the size of my vagina? My current boyfriend says I could be tighter. But my previous boyfriend never had any trouble. Of course, his penis was larger than my current boyfriend's. I'm 28 years old.

You might have your gynecologist check to see if your vaginal opening was overly stretched during childbirth, or if the episiotomy repair was improperly done. If this or other types of vaginal surgery left you with a stretched vagina, corrective surgery may be used to tighten it. But if you don't have such a problem, you're unlikely to find a reputable gynecologist who will perform surgery merely to accommodate a particular boyfriend.

Chances are that your vagina is not too loose, but that its muscles are unexercised and weak. Vaginal exercises called Kegels can improve gripping of the penis and increase enjoyment for both partners. Try this exercise: Alternately squeeze and relax your pelvic muscles as if you were starting and stopping the flow of urine.

A looser feeling inside the vagina also has a great deal to do with a woman's stage of sexual arousal. During the excitement stage, the inner two-thirds of the vagina begin to expand. If penile entry is made during this phase, the woman may feel as if the penis does not fill her vagina. The man may get a similar feeling of looseness.

But as the woman becomes more sexually excited, the outer third of the vagina contracts, gripping the penis. Then the loose feeling disappears for both partners. Thus, the whole problem may be solved by making sure the woman is sufficiently aroused before penetration.

Your current partner's age may be a factor, too. If he's a man in his fifties or sixties, his erections may be less firm. That may account for his feeling that your vagina could be tighter.

If this is the case, you might try some intercourse positions that increase penile friction and give a feeling of greater tightness. In one such position, the woman lies on her back with her legs together. The man lies on top with his legs apart, or with one leg pressed between her legs.

You might experiment with other positions to see which produce the desired sensation of tightness.

I'm a 14-year-old boy, and my breasts are larger than the breasts of some of the girls in my class. The other boys tease me and call me Dolly Parton.

I make excuses to miss gym so I won't have to undress in front of anyone. What can I do?

Many boys your age experience a small swelling of the breasts. It's called gynecomastia, and it occurs in about 60 to 70 percent of normal boys during puberty. Gynecomastia usually develops around the age of 13, and disappears after twelve to eighteen months.

Many boys feel the enlargement as a distinct, tender mass of tissue. It's usually located just beneath the nipple area, and it may occur on one side only. Sometimes the enlargement is so small that it's hardly noticeable.

A small number of boys develop breasts much like those of a girl. Like you, they suffer intense embarrassment. Some become socially withdrawn. Being overweight exaggerates the breast swelling. If that's the case, losing weight (or growing taller) may solve your problem.

It's best to consult a physician so that he or she can rule out endocrine disturbances, sexual disorders, malnutrition, and other problems that could cause breast enlargement. You particularly need to see a physician if the condition developed before any other sign of puberty.

When a boy has had pendulous breasts for three years or more, plastic surgery may be advised. The cosmetic results are usually excellent. But, as a general rule, you can expect your problem to clear up of its own accord.

☐ *Lately, intercourse with my husband of thirty years has become painful. I'm 55 years old and have gone through menopause.*
What could be causing the pain?

One likely possibility is that you're suffering from atrophic vaginitis. This is a degeneration of the vaginal lining caused by the decreased estrogen production that accompanies menopause. Vaginal lubrication diminishes, and a longer period of sexual stimulation may be needed for the vaginal walls to moisten.

Further, with aging, the wall of the vagina loses some elasticity, and the vaginal barrel shrinks. The friction of intercourse may cause the vagina to crack and bleed.

Your physician may prescribe estrogen cream or suppositories to be applied to the vagina. Taking estrogen in this manner causes fewer side effects than taking it in the form of pills or through injections. However, even topical estrogens can be absorbed into the bloodstream, so they should be used cautiously. Estrogens should never be used by women with estrogen-dependent malignancies such as breast or endometrial cancer.

☐ *I've heard that HTLV-III, the virus thought to be responsible for AIDS, can sometimes be found in the saliva of victims. Does this mean that AIDS can be spread through such "casual contact" as kissing?*

There are still many unanswered questions about AIDS, although it is now certain that it's caused by a viral infection. Apparently the HTLV-III virus must enter the bloodstream in order to set up an infection. The virus can enter the bloodstream through blood transfusions and shared drug injection needles, and by crossing the placenta in newborns.

During anal (receptive) intercourse, semen containing the virus can enter the bloodstream through tiny tears in the rectal mucosa. During vaginal intercourse, it can enter the bloodstream through the vascular uterine wall. Although other routes of sexual transmission are possible, passive anal intercourse in gay populations is the strongest epidemiological link.

Now that testing for HTLV-III is possible, many body fluids are being examined for its presence.

Recent evidence, based on human and animal studies, shows that saliva contains the virus, and it may be one mode of transmission. But, says Dr. Robert C. Gallo, a leading AIDS researcher, "right now epidemiological studies do not point to saliva as the key mode of spread of AIDS . . . I don't think saliva is a major route of transmission of AIDS in humans."

Even when we do know all the body fluids that contain the virus, the story won't be complete. We must also know what the body does when it encounters these fluids. For example, even if saliva does contain the virus, it may not be infectious unless it repeatedly comes in contact with an open bleeding area.

Until more is known, preventive measures are based on common sense and careful guesswork. Casual contact, including kissing, has not been linked as yet with the spread of AIDS. Logically, it would seem that if the disease were commonly transmitted this way, AIDS would be much more widespread. However, until more is known about AIDS transmission, deep kissing should be avoided.

I've always understood that a simultaneous orgasm is the highest goal in lovemaking. But my girlfriend doesn't agree. She says they're mostly not worth the trouble. What do you say?

Most authorities agree with your girlfriend. If simultaneous orgasms happen by accident—wonderful. Otherwise, forget them. Far from being a sign of superior sexual achievement, the effort to "come together" can ruin sex.

"Couples may become obsessed with the desirability of mutual orgasm," warns Dr. Benjamin A. Kogan of California State University, Northridge. "In this way, they place one another on trial. That could not only block any hope for a mutual orgasm, it could interfere with the orgasmic process itself."

Striving to coordinate such basically involuntary responses as those leading to orgasm can cause the partners to observe themselves ("Should I wait? Hurry up? Am I too slow? Too fast?"). This kind of spectator role is just the opposite of being immersed in the pleasures of lovemaking. It can lead to impotence and orgasmic difficulties.

What's more, if partners do achieve simultaneous orgasm, they can be deprived of a greater pleasure: observing, without distraction, the other's orgasm.

In general, the most satisfying lovemaking is free of self-consciousness. If you tune in to your body, moment by moment, it's likely to signal what will feel good. Often, these signals will culminate in orgasm. But when the mind superimposes the "goal" of mutual orgasm on the body, the orgasmic process can become thwarted.

___

I'm 19 and pregnant. Where can I go for an abortion where I can feel safe—and not pay a mint?

Planned Parenthood (212-541-7800) has affiliates in most states, and can help you obtain a safe, low-cost abortion. It's a nonprofit agency and charges no fee for referrals. Look for a local listing before you call the national office in New York City.

Since the legalization of abortion in 1973, there has been a proliferation of free-standing abortion clinics. Often you may find them advertised in the yellow pages of your phone directory. Such clinics are usually your best bet for safe, low-cost abortions. Be prepared to spend more money if you're in the second trimester of pregnancy, or if the abortion is performed in a hospital.

It's best to check out the clinic beforehand. When you visit, make sure it looks clean to you. Also look for private space set aside for interviewing and counseling. Are the personnel friendly and non-judgmental? The staff can make a big difference, not only medically, but emotionally as well.

Make sure the facility has arrangements for emergency care should that become necessary because of complications. Regular follow-up care and counseling on birth control should also be available.

Remember to check out any private health insurance your family may have. Some policies pay some or all of abortion costs.

I'm a 14-year-old guy. I'm very embarrassed by erections that just happen when I'm in school. I can simply be sitting there, and suddenly I'll get an erection.

It makes me terribly self-conscious, having this obvious bulge in my pants. What can I do?

Spontaneous erections are quite common during adolescence. They'll become less frequent as you approach your twenties.

In the meantime, says Randall Rissman, a family physician, you should realize that what seems like a great bulge to you is largely invisible to other people. "It's like having a fat lip," said Dr. Rissman. "A little swelling may feel very big and obvious to you, but other people may not even notice."

Since you know you're prone to spontaneous erections, dress so as to conceal them as much as possible. The best combination for concealment may be brief-style shorts (which will hold the erection tighter against your body) with loose pants over them. Loose sweat pants are also likely to conceal an unwanted erection, and may be acceptable as school wear.

My husband and I are very allergic people. We both take shots for allergies, and my husband takes antihistamine medication daily.

What effect might this have on our sex lives? Could it cause impotence or birth defects?

Antihistamines often have a sedative effect that can depress sexual desire. In men, this may lead to impotence. In women, antihistamines can impair vaginal lubrication.

If your husband does experience impotence, he should consult his doctor. He might be able to avoid the problem by taking the drug after intercourse. A modified dosage or a switch to another product could also prove helpful.

Women whose vaginal lubrication decreases because of antihistamines should try a contraceptive jelly or a water-soluble lubricant.

The safety of antihistamines during pregnancy and lactation hasn't been established, so the drugs shouldn't be used by women who are pregnant or nursing.

My menstrual cycle lasts forty days instead of the usual twenty-eight. Does this longer cycle reduce my ability to conceive?

Yes, your longer cycle reduces your chance of conceiving.

The middle four days of any menstrual cycle are the best days for conception. But the longer the cycle, the fewer such days there are in a woman's life. During one year, for example, a woman on a twenty-eight-day cycle experiences thirteen full menstrual periods, and a total of fifty-two fertile days.

In contrast, a woman with a forty-eight-day cycle has only nine full periods and thirty-six fertile days.

What's the most widely used form of contraception?

According to a recent study, the most widely used birth control method among both men and women is sterilization.

A report by the National Center for Health Statistics found that 18 percent of couples with partners aged 15 to 44 avoided pregnancy through sterilization of either partner. (Sterilization procedures are vasectomy for the male and tubal sterilization for the female.) In contrast, 16 percent of the couples used birth control pills, 7 percent used condoms, 5 percent used diaphragms, and 4 percent used IUDs. The remainder are unspecified or use no contraception.

Sterilization is particularly popular among married couples who consider their families complete. Among such couples, the proportion using male or female sterilization more than tripled between 1965 and 1982, from 18 to 62 percent.

Among couples who want more children later, the Pill remains the most popular method of birth control—although many more couples are using the diaphragm than previously. Birth control pills were used by 23.9 percent of married couples in 1965, 36.1 percent in 1973, and 19.8 percent in 1982, according to the report.

Female sterilization has increased dramatically. From 7 percent in 1965, it grew to 12 percent by 1973. By 1982, it had more than doubled to 26 percent. Male sterilization rose from 5 percent in 1965 to over 15 percent in 1982. When women are considered separately, though, more of them use the Pill than are sterilized. The 8.4 million women who use birth control pills outnumber the 6.4 million women who have been sterilized.

But over 3 million men have been sterilized. Together, the nearly 10 million men and women who have been sterilized make sterilization the most common birth control method.

I'm a 60-year-old male. I have severe arthritis in my hands and for some time I've been unable to masturbate. I used to enjoy this activity very much. Do you have any suggestions?

Yes, we've heard of a device called Accu Jac. It's apparently a mechanical masturbator. A man puts his erect penis inside. The machine then operates like an electrical cow milker, enabling a man to reach orgasm and ejaculate. This and other such masturbating devices are available from sexual aids catalogs.

It seems especially suited to handicapped men with poor control of their hands.

I've heard about a natural Chinese contraceptive for males made out of cottonseed. Can we get it in this country? Is it safe and effective?

You're thinking of gossypol, an extract from the seed of the cotton plant.

The Chinese claim a 99.8 percent rate of effectiveness for the contraceptive. Men in China's Hebei Province who were taking gossypol showed a very low rate of fertility. Apparently, gossypol works by interfering with sperm maturation and motility. American scientists discovered that it inhibits an enzyme that is essential to the function of sperm.

In its current form, however, gossypol is not considered a viable male contraceptive because of its serious side effects. Some Chinese men taking gossypol suffered heart irregularity. Cases of cardiac arrest and death were also reported. Another drawback is that gossypol, a fatty substance, may accumulate in tissue and cause circulatory problems. It can also cause loss of appetite, anemia, reduction in the growth rate, vomiting, and diarrhea—even if taken in low doses.

It's ironic that gossypol, while providing effective contraception, also caused impotence in some users. So, all in all, gossypol is not the "miracle" male contraceptive we've been waiting for.

I've read that males cannot have multiple orgasms in the same way that women can. But I can sometimes have as many as four peaks, one right after the other. I'll seem to be sliding down from an orgasmic peak, and then I'll head right back up again and have another orgasm that's just as strong.

I've had these peaks for most of my sexually active life (I'm 31 now). Do you agree that they're multiple orgasms?

According to Richard Reznichek, M.D., and Carol Reznichek, R.N., a sex therapy team, multiple orgasm in men can include just what you describe.

Many people confuse orgasm and ejaculation, but they are not identical. A man can ejaculate only once in the course of each sex act, but he can have several orgasms before ejaculating.

In addition, some young men are fortunate enough to have very brief periods of resolution between sexual episodes. That means they can quickly achieve another erection and, thus, can ejaculate several times in a brief time period. The period of resolution—the time between erections—tends to lengthen with age.

 Is it true that most rapes are committed under the influence of alcohol?

Studies show that about half of sexual offenses are alcohol-related.

In research conducted by Elaine P. Bencivengo and Joseph J. Romero of the Joseph J. Peters Institute in Philadelphia, about 53 percent of sex offenders reported having been drinking when they committed the offense.

Studies by Richard T. Rada, M.D., confirm these findings. Fifty percent of the rapists interviewed for his book, *Clinical Aspects of the Rapist*, said they had been drinking before the crime.

"It is clear that drinking is frequently associated with the commission of a sex offense," say Bencivengo and Romero, "especially if there is a history of alcoholism." But they point out that most researchers do not consider alcohol use to be the primary reason for the sex offender's behavior.

■ *I know you're not a virgin if you've had intercourse. But what if you've had oral sex? Or anal sex?*

And what about gays? Or people who do heavy petting and have orgasms? To sum up, when are you not a virgin any more? (I'm 14, and a virgin no matter how you define it.)

You ask a very good question—and one that points up how hard "virginity" is to define.

Conventionally, virginity is the state of not having had sexual intercourse. And sexual intercourse is usually understood to mean heterosexual penile-vaginal penetration. But researchers are finding that this may be an outmoded definition.

In a study of 312 college students, almost half felt that female virginity was a meaningless or unimportant concept. Of those who considered virginity a valid concept, 81 percent said that full penetration of the vagina by a penis would constitute loss of virginity. But some of the students suggested other possibilities that would also qualify as losing one's virginity. Among them:

● A woman's bringing herself to orgasm.

- Rupturing of the hymen.
- A male's bringing the woman to orgasm.
- Partial penetration of the vagina by the penis.
- Penetration of the vagina by something other than a penis.

Many people would add oral and anal sexual activity—either heterosexual or homosexual—to the list of activities that constitute loss of virginity for either sex.

I'm 41, and I've been considering having a vasectomy.

However, my girl, who is into oral sex, is worried. She thinks that not having enough sperm cells swimming around will change the consistency of the semen. She says that the sperm give the semen a thick base. Without them, it will be very thin.

Is this true?

Sperm constitutes a very small fraction of semen, only about 1 percent. Most of the semen (about 95 percent) is made up of fluids secreted by the prostate gland.

After vasectomy, there is no noticeable change in the volume or consistency of the semen. It's very unlikely that your girl will be able to perceive any difference.

I'm 19. Shortly after first having intercourse a few weeks ago, the head of my penis began to blister and peel. I went to the VD clinic right away, but they couldn't find anything wrong.

What could have caused this?

Armando DeMoya, M.D., and Dorothy DeMoya, R.N., M.S.N., speculate that your reaction may have been caused by allergy.

Did you wear a condom? "In our experience," say the DeMoyas, "the most common cause of the condition you describe is hypersensitivity to the rubber or lubricant on standard condoms. The difficulty can be prevented by using a condom made of lambskin, which is hypoallergenic."

If you didn't wear a condom, you may be allergic to chemicals in vaginal contraceptives and douches, which your partner should thus avoid.

Or you may be hypersensitive to vaginal secretions. In that case, wearing a condom would reduce your chances of having an allergic reaction, although it may be sufficient to wash immediately after intercourse. You may find, by the way, that you are allergic to the secretions of some women but not others.

My 20-year-old brother was recently injured in a motorcycle accident. His spinal cord was damaged, and his legs are paralyzed.

I'm wondering what his sex life will be like.

Spinal-cord injury has varying effects on sexual functioning. No absolute predictions can be made about the extent of sexual response that will be retained after a particular injury.

That's why patience is a prime requisite for people with spinal-cord injury. Sometimes, sexual response and enjoyment improve with time. Occasionally, sexual functioning returns as long as two or three years after the injury.

Even with impaired sexual functioning, however, many spinal-cord injured men experience pleasurable sensations. Physical contact can provide a sense of intimacy and well-being.

Men with spinal-cord injury usually retain sexual desire, but it may be diminished by such factors as depression and anxiety.

While erection, ejaculation, and orgasm may occur after spinal-cord injury, most men suffer some degree of impaired sexual functioning. For example, only 25 percent of spinal-cord injured men can maintain an erection that is sufficient for intercourse. Even those 25 percent have intercourse considerably less often than before the injury.

The ability to ejaculate is lost by most spinal-cord injured men, meaning that fertility is greatly reduced. Those men who do ejaculate may not have the sensation of orgasm.

On the other hand, some men experience "phantom orgasm." Using erotic imagery, they can have imaginary sexual experiences that culminate in orgasm. Such orgasms show none of the physiological changes that generally accompany orgasm. Conversely, some men undergo the biological changes of arousal and orgasm, but do not have any sensation.

Rehabilitation programs in hospitals that treat spinal-cord injury help individuals become more self-reliant. Many provide individual psychological counseling or group-therapy programs. Some facilities also have special sex counseling for patients.

The following publications may be helpful to your brother: *The Sensuous Wheeler*, by Barry J. Rabin (MultiMedia Resource Center, 1525 Franklin Street, San Francisco, CA 94109); and *Sex and the Spinal Cord Injured: Some Questions and Answers*, by M. G. Eisenberg and L. C. Rustad (U.S. Government Printing Office, Washington, DC).

■ *I have a secret passion to make love to a corpse instead of my girlfriend. She does not know about this obsession that is constantly on my mind.*

Is this a normal thing to think about? I'm 18 years old.

No, it's not. Harold I. Lief, a psychiatrist, explains that a secret desire to make love to a corpse is called necrophilia. In his view, it indicates a very severe disturbance, and he advises you to seek psychiatric care.

Although necrophilia is a rare sexual deviation that has been poorly studied, most experts agree with Dr. Lief that it signifies a profound emotional disturbance.

In your case, necrophilia is a fantasy, an impulse that is in your conscious mind, but not yet acted upon. The fact that you're worried about having such thoughts indicates a desire to seek help.

Psychiatric help may give you insight into the origins of such fantasies and desires. One researcher suggests that necrophilia may be explained as an attempt by the individual to dominate someone, even if it is a cadaver.

Some people who practice necrophilia are psychotic. Their sexual gratification may come from looking at a corpse, having intercourse with it, or mutilating it.

Reports Dr. James Leslie McCary, co-author of *McCary's Human Sexuality*, a college textbook: "The necrophile may kill to provide himself with a corpse, have sexual relations with it, mutilate it, and even cannibalize it. Most professionals view necrophilia as the most deviant of all sexual aberrations.''

I'm never quite sure that my girlfriend has really had an orgasm.
She seems to. But I'm wondering if there's a way to tell for sure, by physical signs.

''No, there is no infallible way of telling if a partner has really had an orgasm,'' responds Dr. Dolores Elaine Weller, professor of biology and psychology at Pace University in New York.

''A woman who wants to fake orgasm can usually do a convincing job. Aside from being skilled at acting the behavioral components of orgasm, including moaning and thrashing, many informed women can also simulate the physiologic components by contracting their vaginal muscles at approximately 0.8-second intervals.''

There's a myth that nipple erection in women is a sign that they've had an orgasm. Actually, breast changes accompany sexual arousal, which may or may not be followed by orgasm. What's more, the changes may not even be visible.

If you suspect that your girlfriend is faking orgasm, you may want to bring up the subject of mutual sexual satisfaction. Ask her if she's happy about your sex life. Ask her what she particularly likes or dislikes. If you communicate well, she should not feel the need to fake an orgasm.

I'm a 26-year-old woman who recently lost thirty pounds. To show off my new figure, I've been wearing very tight jeans. But I've also developed vaginal infections. My doctor says they come from fecal contamination. He thinks the jeans are at fault. Is that possible?

Yes. Wearing very tight jeans promotes sweating of the perineum, the area between the vagina and the anus. If intestinal microorganisms are contaminating the anal region, the sweating promotes their migration to the vagina. One way to prevent infections that are caused in this way is to wear jeans that are looser around the seat.

I've heard that Anne Boleyn, a wife of Henry VIII, had three breasts.

Is this true? How common is it for women to have extra breasts?

That was certainly the rumor about Anne Boleyn, but we don't know if it's true. However, extra breast tissue and extra nipples "are quite common," according to Takey Crist, M.D., assistant clinical professor of obstetrics and gynecology at the University of North Carolina.

It's estimated that 1 to 3 percent of healthy normal people—both men and women—have extra breast tissue. Such breast tissue may occur with or without a nipple. The tissue is usually found below and between the normal breasts. Sometimes it appears under the arms.

If there's a nipple present, says Dr. Crist, it can even secrete milk after childbirth.

If the breast tissue has no nipple, it may often go undetected, or it may be mistaken for other kinds of growth. In addition, breast tissue that consists of only a small nipple and areola may often be mistaken for a mole or birthmark, and may not trouble the person.

Extra breast tissue is subject to the maladies of other breast tissue: infection, abscess, and fibrocystic breast disease. It may also respond to hormones during menstruation, pregnancy, and after childbirth. But extra tissue is not usually capable of erotic sensations.

Abnormal breast tissue usually can be surgically removed, and many people elect to do this for cosmetic reasons. It's a good idea to remove extra tissue if it begins to grow or cause pain.

In rare cases, extra breasts are fully developed, with underlying breast tissue as well as a nipple and areola. The extra breast is located just below a normal one. It looks as if two small breasts have been incorporated into a larger one with two nipples.

Such complete breasts are likely to be erotically responsive. Most can be surgically removed with good cosmetic results.

I would like information on arthritis and sexual relations.

I am a 24-year-old woman who has Lyme disease, and I have developed an arthritic condition. It's being treated with nonsteroidal anti-inflammatory drugs.

Your answer will help me and probably some of the older subscribers as well.

Arthritis (chronic joint inflammation) can hamper sexual activity because of stiffness and pain.

To help avert these problems, it's best to plan intercourse for the time of day when pain and stiffness are usually least severe. Many patients find that stiffness is greatest in the morning and subsides somewhat with moderate activity. Thus, early afternoon to evening may be the best time for sex.

You can also prepare for intercourse by taking your pain and anti-inflammatory medications before sexual activity. Loosening-up exercises prior to intercourse may also help avert pain and increase your range of motion.

You might also take a warm bath or shower beforehand, or apply hot packs to afflicted areas, since wet heat often brings relief. If your partner is agreeable, you can even make the bathing part of your lovemaking.

Experiment with a variety of positions to discover which ones are the most comfortable. One problem is likely to be keeping weight off the hip and spine. The female-superior (on top) position usually puts the least amount of strain on an arthritic woman. The couple may also enjoy positions that permit vaginal entry from the rear, or allow both partners to lie on their sides. A water bed may help reduce the fatigue that can come from sexual activity.

Many experts advise that sex is good for arthritics. It can have a painkilling effect that sometimes lasts for hours.

A friend of mine, who's 22 years old, had both his testicles crushed in an accident, and they had to be surgically removed.
Will he still be able to function sexually?

In all likelihood, your friend will still be capable of functioning sexually.

In cases of accidental castration like his, doctors usually prescribe the male hormone testosterone, which is provided mainly by the testicles. The prescribed testosterone is generally given either by injection or by placing a tablet under the tongue.

Testosterone will protect your friend from becoming impotent. It will also preserve such other testosterone-dependent functions as beard growth, production of prostatic secretions, and maintenance of muscle mass.

Thus, he'll physiologically be able to have intercourse and ejaculate. But he won't be able to have children, since without testicles he won't produce sperm.

I'm a 17-year-old boy. Some years ago my dad and I used to play spanking games. Now I can only have an orgasm when I think of spanking boys.
Where can I get help?

Your question indicates that you realize this is a serious problem.

Richard V. Lee, M.D., a pediatric internist, recommends that you first look for help from your doctor. If he or she feels uncomfortable with the subject, or unequipped to deal with it, you'll be referred to a specialist.

Planned Parenthood clinics, school counselors, and family service agencies can also be a source of advice and referral.

Regardless of the route you choose, make sure that the person who counsels you can help you resolve issues of sexual identity. You should also recognize that there is no easy ''cure'' for the concerns you have. Working with these kinds of problems takes time and great effort.

I read that a woman was arrested for sodomizing her son. Can this be?

Legally, sodomy is defined differently in different states. Usually it means anal sex. But sometimes it means oral sex as well. So the woman you read about had probably performed oral sex on her son.

I enjoy oral sex, and I like swallowing my husband's semen. Is there any danger in doing this?

Swallowing semen is generally harmless.

Ingesting it has not been known to cause food-type allergic reactions. However, if the semen is contaminated with microorganisms, disease could result.

Gonorrhea or syphilis of the throat may be caused by contaminated semen. Viral disease, including herpes simplex, AIDS, and cytomegalovirus infections, can also be transmitted this way.

Can you describe how test tube babies are made?

Sex educator Judy Henkel notes that the expression "test tube baby" is misleading, as these babies are not grown in test tubes. They are, however, conceived outside of the mother's body by fertilizing an egg in vitro, meaning "in glass." The technique is most often used for women who are unable to conceive because of blocked fallopian tubes.

For the procedure to work, the woman must ovulate (release an egg from her ovary) each month. Through careful monitoring of her hormone levels, her physician is able to determine the appropriate time to remove the ripened egg.

Using a laparoscope, an instrument designed to illuminate and magnify the inside of the abdomen, the physician can locate the egg and remove it with a special syringe. The egg is then placed in a sterile petri dish that has been prepared with a solution that resembles the conditions in the fallopian tube. Sperm are then added to the dish.

The petri dish is incubated for two to four days. If the procedure is successful, the egg will be fertilized and cell division will proceed as it does in a normal pregnancy. By the fifth day the sixteen-cell embryo is ready to be implanted in the woman's uterine wall.

For this part of the procedure, the physician draws the tiny mass of cells from the petri dish into special tubing that is inserted through the vagina and cervix into the uterus. Here it is expelled, to implant in the uterine wall and grow to term, nourished by the mother.

Sometimes after my girlfriend and I have fooled around but didn't have sex, I get an ache in my testicles. I say I've got "blue balls." She says that's a myth. Who's right?

You are. "Blue balls" is a slang term for a condition doctors call epididymo-orchitis. The term describes inflammation of the testicles and the epididymis, an oblong organ attached to the testicles.

Blood vessel congestion from ungratified sexual excitement, such as you describe, can sometimes cause an ache in the testicles. It's not a serious problem, and does not require a doctor's care. Teenagers are especially susceptible to it because of their dating patterns and because they are at a stage of heightened physical responsiveness.

Boys may fear they will suffer permanent damage. In fact, the condition is self-limiting. It will usually go away by itself within a few hours at most.

Some boys obtain relief through masturbation.

Why is it that I get most interested in sex just when I'm starting my period?

Many women share your experience. Heightened sexual desire as menstruation begins has to do with increased blood flow in the pelvic area. The vasocongestion engorges the vaginal lips, increases sensitivity of the vulva, and heightens feelings of arousal.

My girlfriend sometimes seems to have orgasms through breast fondling alone. Is this possible, or is she putting me on?

Yes, it is possible for some women to achieve orgasm solely through breast stimulation.

Masters and Johnson and other sex researchers have reported this response. In their original research group, Masters and Johnson studied three women who were able to achieve orgasmic response by breast fondling alone. An interesting finding was that the same genital changes occur whether an orgasm is achieved by breast fondling or intercourse.

Some women report reaching orgasm without any tactile stimulation at all—through fantasy or reading erotic literature.

I know that oral acyclovir (Zovirax) has been approved by the Food and Drug Administration for the treatment of genital herpes.
Can I expect any side effects if I take the drug?

The manufacturer of Zovirax (Burroughs Wellcome Co.) reports that the most frequent adverse reactions in clinical trials during short-term administration were nausea and/or vomiting (in 8 of 298 patients) and headache (in 2 of 298 patients).

In long-term administration of the drug (daily therapy for three to six months), 33 of 251 patients experienced headache, 22 had diarrhea, and 20 reported nausea and/or vomiting. Nine of the 251 patients suffered vertigo, and another 9 had joint pain.

Zovirax did not cause birth defects in animal studies. But, because there have been no well-controlled studies of the effect of the drug on pregnant women, it should not routinely be given to them. Nursing mothers should avoid Zovirax, too.

Zovirax capsules are intended for the treatment of initial episodes and the management of recurrent episodes of genital herpes. The drug is mainly used for patients who have severe and frequent outbreaks. While Zovirax can suppress symptoms and recurrent outbreaks in some patients, it's not a cure for herpes.

The safety and effectiveness of oral acyclovir has been established for only up to six months. The manufacturer recommends that it be used only for that amount of time until further studies have been done.

Some questions about the safety of acyclovir remain. High concentrations of the drug caused decreased sperm production and chromosomal changes in animals. That's one reason why the recommended dosage or length of treatment should not be exceeded.

It's also possible that resistant forms of the virus may emerge with repeated or prolonged courses of acyclovir therapy. These resistant viruses may not fully respond to continued therapy.

If I masturbate every day, or have sex frequently, will I run out of sperm?

This is a common misconception.

Many men believe they are allotted a certain number of sperm. They fear that if they expend the supply through masturbation or frequent intercourse, they won't have enough left to father a child later on.

In fact, a male's testicles keep on producing sperm from puberty until death, although after 70 or so there is usually some decrease in the amount. A healthy male produces about 72 million new sperm each day.

So even if a man ejaculates frequently, his semen will still contain many millions of sperm.

If my girlfriend doesn't have an orgasm when we have intercourse, does that mean I'm a bad lover? We're both 18.

The assumption underlying your question is that in lovemaking the male is responsible for the female's satisfaction. This places a great burden on males. If the man does not receive sexual satisfaction, it's considered his own fault. If his partner is not satisfied, that's his fault as well.

Increasingly, however, sex counselors and therapists feel that each partner in a sexual relationship must accept responsibility for his or her own sexual satisfaction.

Sexual communication is the cornerstone of a good sexual relationship. Don't assume that you will intuitively know what your girlfriend likes and dislikes. Instead, ask her to tell you her sexual preferences.

Many women find, by the way, that they can gain useful knowledge about what they enjoy sexually through masturbation.

Bear in mind that most women do not reach orgasm through the thrusting of intercourse alone. Research shows that roughly two out of three women require more direct clitoral stimulation than penile thrusting can provide. If your girlfriend asks you for such manual stimulation, comply. Meeting a partner's needs is the mark of a good lover.

In high school, my sex ed teacher said that urine kills sperm. He said that for ladies, urinating after sex is a somewhat effective form of birth control.
Is this true?

"Some people believe that urination immediately following intercourse acts as a contraceptive method," says sex educator Dr. Michael Carrera, author of *Sex: The Facts, the Acts, and Your Feelings.*

This is a myth. "Since urination is through the urethra and not out of the vagina, sperm cannot be flushed out of the vagina in this fashion," he advises. "People who practice this method of birth control are usually called parents."

My 15-year-old niece was just found to have been born without a vagina. What can be done for her? Will she be able to have children?

According to the *American Journal of Diseases of Children*, congenital absence of a vagina is a fairly common problem. It occurs once in every 4,000 to 5,000 female births.

Generally, the clitoris and labia minora and majora are normal. But the uterus is often absent or not fully formed. Absence of a vagina may also be associated with other pelvic and kidney abnormalities.

After surgery to open the vaginal canal, affected youngsters have the capacity to engage in sexual intercourse. In some cases, progressive dilation is used to widen an undeveloped vagina.

When the uterus is normal, surgery can usually be performed to allow fertility. So your niece's ability to bear children depends on whether her uterus is fully developed and functional.

As with your niece, absence of the vagina is not usually discovered until a girl is in her mid-teens and fails to menstruate. The most damaging results of the condition can be psychological—shock, depression, and lowered self-esteem. After treatment, some girls react by becoming sexually promiscuous. Your niece and her family may benefit from counseling to help them anticipate or resolve problems caused by her condition.

To discover and treat the absence of a vagina early, it is wise for a girl to have a complete pelvic examination early in her teens.

I'm afraid my 14-year-old son may become ho-mosexual. He's very attached to a friend two years older, and neither of them seems interested in girls.

Is this normal? Or is he on his way to becoming a homosexual?

There are many theories about the cause of homosexuality, says sex educator Judy Henkel. But none of the numerous studies that have been done have explained conclusively why approximately 10 percent of the population is homosexual.

You should remember that normal sexual development embraces considerable experimentation. This may include homosexual experience. But such experience does not usually mean that a lifetime of homosexual behavior will follow.

There are many reasons why a young adolescent may come to think of himself or herself as homosexual. These include feeling different from peers, having a crush on someone of the same sex, fantasizing about someone of the same sex, not being interested in people of the opposite sex, or having engaged in sporadic homosexual behavior.

None of these factors, however, means that a teenager is predisposed to become a homosexual. If parents are uninformed about normal adolescent development and use the same erroneous criteria, they can draw the wrong conclusions about their child.

I stopped using contraception when I was 49 and was pretty sure I was menopausal. So I was amazed to find myself pregnant (I had a miscarriage).

How long should you keep using birth control? What type is best?

Physicians usually recommend that a menopausal woman use contraception for a year following her last menstrual period. As you discovered, ovulation can occur even when periods are widely spaced and the woman thinks her reproductive life is over.

At the menopausal age, there is a much higher death rate for both fetus and mother than there is during the usual childbearing years.

The best types of birth control for a woman going through menopause are the diaphragm or IUD. The Pill is inadvisable, since it is associated with a greater risk of cardiovascular difficulties and other side effects in older women.

What's the percentage of women in the United States who are still virgins when they marry? It's my impression that most women aren't.

Your impression matches government statistics. According to a report by the National Center for Health Statistics, more than three-quarters of American women have sexual intercourse before marriage.

This represents a sharp increase from twenty-five years ago. Among women who married from 1960 to 1969, almost half—48 percent —delayed sexual intercourse until marriage. In contrast, of women who married between 1975 and 1979, only 21 percent were virgins.

The Center's report was based on interviews with 7,969 women, ages 15 to 44.

I'm a 25-year-old woman. One of my breasts is significantly smaller than the other.

Should I have an implant done to make them equal? What's involved?

It's common for women to have a difference in the size or shape of their breasts. The difference is usually not large enough to cause them any problem.

But a significant size difference may sometimes make a woman feel self-conscious and sexually inadequate. You are evidently feeling uncomfortable enough about the size difference to consider breast augmentation surgery.

To explore such surgery, consult a board-certified plastic surgeon. He or she may advise against the expense and risk of surgery because the difference in breast size is not as great as you perceive it to be. On the other hand, the surgeon may agree that surgery is needed and would give you excellent results.

As with all surgery, it's usually a good idea to get a second opinion before proceeding.

In general, breast augmentation surgery takes about an hour, and can usually be done on an out-patient basis. It is performed by cutting an incision at the underfold of the breast and inserting a silicone-rubber bag filled with silicone gel.

There will be a fairly small scar beneath the breast, and there is usually a degree of discomfort during a two-to-four-week healing process.

Breast augmentation should not interfere with breastfeeding. Breast sensitivity should also not be impaired. An implanted breast may feel somewhat firmer than a normal breast. In general, the more breast tissue that covers the implant, the more natural it will feel.

Complications of breast augmentation are rare. Sometimes breast implants may sag or migrate, causing an awkward look. Some women experience discomfort and drainage from the surgical incision that may persist. In such cases, the implants might have to be removed.

My boyfriend and I (we're both 16) have been using whipped cream as a spermicide, and I haven't become pregnant. Is it safe to continue using this method?

There's no evidence that whipped cream works as a spermicide.

It may look like spermicides such as foam, but it doesn't have foam's sperm-killing chemicals. If you've been using whipped cream as birth control and you haven't become pregnant, you're just lucky.

Could you tell me different ways to masturbate?
For several years, I've been simply stroking my penis with my hand. This is getting boring.

Good for you for recognizing that masturbation is a valid form of sexual activity. When you experiment with masturbatory techniques, you can find out what you like and what you don't like. Thus, not only does masturbating become more enjoyable, but you can also get more out of sex with a partner.

One of the areas to explore is the sensitivity of the penis. The most sensitive part is the frenulum, the small fold of tissue under the glans (head). The corona (rim) of the glans is also extremely sensitive, especially in a man who is not circumcised.

Individual men also find special areas of pleasure in the base of the penis, in the shaft, and in the scrotum. Each of these parts can offer many different experiences.

Many men enjoy testing various pressures and speeds of manipulation. For example, slow squeezing provides one kind of sensation, while hard, vigorous manipulation affords another. Kneading is another enjoyable technique.

Masturbating slowly and delaying ejaculation is not only pleasurable, it's also helpful in learning how to avoid premature ejaculation.

Masturbating can involve much more than the penis. The entire body can be an instrument for pleasure. There's erogenous potential anywhere that mucous membrane meets skin. That happens in the lips, the nostrils, and the anus. In addition, the nipples, ears, and feet can produce heightened sensations.

Stretching and contracting muscles can produce pleasurable sensations. Pressing the buttocks together, for example, may particularly increase genital feeling.

You can intensify sexual sensations by concentrating on breathing techniques such as inhaling deeply and slowly, panting, and gasping. Some men find that humming and other sounds can increase pleasure.

Many men find fantasy—erotic movies of the mind—an aid to masturbation. Remember, though, not to be troubled by whatever the imagination conjures up. The unconscious has a life of its own and is not subject to conventional ideas of right and wrong.

Men may find themselves, for example, thinking of having sex with other males or family members. They may imagine themselves engaging in other acts that they'd ordinarily find objectionable. None of this has anything to do with reality. Many "normal" people have "perverted" thoughts. In sex as in other areas of life, behavior controls almost certainly will keep the fantasizers from acting in an unacceptable manner.

To stimulate fantasy, many people find erotic stories and pictures helpful.

Different types of lubricants may increase pleasure. Men can reduce friction and gain a variety of sensations with cold cream, K-Y Jelly, petroleum jelly (such as Vaseline), baby oil, vegetable oil, lotions, and the like.

It's also fun to experiment with different objects against the skin—feathers, or other soft objects, for example. Some people enjoy rubbing with coarser substances, such as towels and sponges. Vibrators afford a variety of sensations. And music may also enhance the masturbatory experience.

CONFESSIONAL BOX

"I can't get enough of masturbation"

<u>Eva, (33)</u>: I began masturbating at age 9, had my first orgasm at age 11, and was instantly addicted. Within a week I was multiorgasmic. As soon as I found out that the incredible feelings I had were not harmful, I couldn't masturbate enough. I've been doing it daily ever since, sometimes even several times a day. It won't hurt you and can only improve your life.

My husband doesn't know I do it, but he sure loves the way I go crazy, cry out, and carry on while climaxing like a string of firecrackers. He thinks it's him. Well, it is, sort of, but my conditioned body is a big help. I love to come and can't get enough. Anybody who tells you playing with yourself is wrong, is nuts.

Is it possible for a woman to be allergic to a medication in her husband's semen?

Although this intriguing question has not been extensively studied, reports from individual physicians suggest that a woman may indeed have an allergic sensitivity to a medication her husband is taking.

In the February 1985 issue of the *American Journal of Psychiatry*, Dr. M. D. Sell of Augusta, Georgia, described a 28-year-old female patient who had never experienced any sort of allergic reaction. But after her husband began taking the antipsychotic thioridazine (Mellaril), the woman developed an itching rash on her genitals. Clinical tests showed that she was allergic to thioridazine. After her husband began using a condom, the problem cleared up.

Another physician reported vaginitis in a woman whose husband was taking medication for Hodgkin's disease.

My wife has just been diagnosed as having multiple sclerosis. We're both wondering what effect her illness may have on our sex life.

Multiple sclerosis is a neurological disorder that typically strikes young adults between 20 and 40. It's characterized by periods of exacerbation and remission that leave cumulative damage. Over many years, the victim usually becomes progressively more disabled.

Since the course of the disease is so variable, it's very hard to predict its effect on sexual functioning. Your wife, we hope, will have a very moderate, slowly progressing form of the illness. She may exhibit no sexual problems at all. But, unfortunately, sexual dysfunction is often a prominent early symptom of the disorder. At first, the dysfunction clears up when the disease is in remission. As the disease progresses, however, the sexual impairment is likely to become permanent.

The impairment may take a number of forms. It may involve a decreased interest in sex and problems in attaining orgasm. Vaginal dryness may also be a problem. If this is the case, a woman can use a lubricant such as K-Y Jelly.

Sexual activity can also be difficult because of motor disturbances, such as weakness, difficulty in movement, and spasms. Similarly, sensory disturbances may interfere with sexual functioning.

Genital numbness is also common. A woman may have no feeling in her vagina or clitoris. Or there may be too much sensitivity in the genitals, which makes sexual activity unpleasant. If the over-sensitive area is in the external genitals, applying a topical anesthetic cream before intercourse can relieve the problem.

Psychological disturbances may also contribute to sexual dysfunction. Most people mistakenly assume that a diagnosis of multiple sclerosis always means eventual confinement in a wheelchair. Anxiety, depression, and fear are common reactions, and they usually interfere with sexual functioning.

Many multiple sclerosis patients benefit from group counseling. Therapists can help patients and their families adjust to the uneven course of the disease and cope with progressive disability.

My girlfriend and I have been going together for two years (we're both 17). We've decided to have intercourse but want to use good contraception.
Can minors legally buy contraceptives at a pharmacy?

A Supreme Court decision has made it legal for minors to buy over-the-counter contraceptives (contraceptives that do not require a doctor's prescription) in every state.

Such contraceptives include condoms and spermicides, such as foams, creams, jellies, foaming tablets, and suppositories. In contrast, prescription contraceptives include the Pill, IUD, and diaphragm.

The most effective contraception you can arrange without a prescription is to purchase both foam and a condom and use them together.

Do males have a menopause like women do?

There is no male menopause in the same sense that women experience menopause. The male does not lose his reproductive function, even though production of testosterone, the male hormone, may decrease slightly in midlife.

What's called ''male menopause'' may relate to the fact that the natural aging process causes a somewhat lowered sex drive and decreased potency (a lessened ability to have and maintain an erection). That change in sexual functioning can contribute to self-doubt and depression in some males. It should be noted, however, that poor health, drugs, some medications, and psychological pressures may also be responsible for loss of sex drive.

I'm 31 years old and about to have a vasectomy. Will I have the same level of sexual ability after the operation?

Vasectomy generally leaves sexual performance unimpaired.

This surgical operation for keeping sperm out of the semen is an increasingly common form of birth control. An estimated 1 million men a year have vasectomies.

The glands that manufacture semen are not affected by the operation. You can continue to have intercourse, ejaculate normally, and enjoy orgasm as before. The vasectomy has no effect on the production of the male sex hormones. Nor does it alter the appearance of the genitals.

Vasectomy creates only one physical change: There are no longer sperm in the semen, and hence no possibility of pregnancy. The testicles continue to produce sperm cells but the cells have no outlet. Thus, they dissolve and are harmlessly absorbed into the bloodstream.

You may have intercourse as soon as you desire after vasectomy. Many men find that sex is comfortable after ten to fourteen days. Some are able to resume much sooner.

It's important to use contraception until your doctor has examined your semen and found it free of sperm. Urologist Sumner Marshall of the University of California School of Medicine in San Francisco considers his postvasectomy patients "safe" after a minimum of two consecutive sperm-free specimens. The last specimen must have been produced at least four months after the vasectomy.

Couples often report that they enjoy sex more after vasectomy since they feel confident there will not be a pregnancy. According to Dr. Helen Edey, a member of the medical committee of the Association for Voluntary Sterilization, about two-thirds of vasectomized men report an increase in sexual pleasure during intercourse.

On the other hand, for some men, vasectomy triggers emotional problems. If a man believes his virility depends on the ability to produce a child, he may become impotent.

Some men react to vasectomy with greatly increased, sometimes compulsive, sexual activity. For example, a husband may want sexual relations up to three or four times a day—considerably more often than was usual for him. This compulsiveness can last for fifteen months or so after the operation. Or, as compensation for what he regards as lost masculinity, a man may refuse to do "women's work," like washing the dishes or helping with the children.

Psychological disturbances are most common in men who already had doubts about their sexual adequacy or who experience serious conflicts with their sex partners. If a man hoped, unrealistically, that the vasectomy would solve sexual and marital problems, the disappointment may be upsetting.

A man is particularly likely to have emotional difficulties if he submitted to the operation under pressure from his wife. The opposite is also true. He may have a negative reaction if he had the vasectomy against his wife's wishes. A man should not consider vasectomy if such conflicts exist. They increase the likelihood that he will suffer psychological problems afterward.

How can I handle the stress of worrying about getting AIDS?

I'm 26 years old and gay. Every day—no, every minute—I'm engulfed by the fear that I've contracted AIDS and that I'll soon come down with it. Sometimes I get so depressed about it that I do something stupid, like taking a risk with some stranger. It seems that people who think they might get AIDS are in worse mental shape than those who've already gotten it.

Fred Westendarp, M.D., points out that worry is seldom helpful, unless it's translated into effective action. To take action, have your blood tested for the presence of the HTLV-III antibody. Testing is being done in Alternative Test Centers across the country. There's probably a center near you.

However, you may have to travel out-of-state to find a center that performs anonymous testing. The San Francisco Department of Public Health is one such center, and the testing is free (415-621-4858). For information about centers on the East Coast, contact the Gay Men's Health Crisis AIDS Hotline (212-807-6655).

A negative test is probably good evidence that you have not been infected by the virus. You should have the test repeated in three months, however, just to be sure. In very early infections, the antibody may not be detectable.

Naturally, if you test negative, you'll want to remain that way. Here are some tips for avoiding AIDS:

• Avoid contact with blood—you should not share razors or toothbrushes.

• Practice safe sex at all times. That includes massage, hugging, mutual masturbation, dry kissing, body-to-body rubbing, and the sharing of fantasies.

• Avoid the exchange of body fluids. The most risky thing you can do is to engage in anal intercourse. Swallowing a partner's semen is a close second. It may help somewhat to use condoms consistently if you do have anal intercourse, but the risk is still substantial.

The following practices rank in the "possibly safe" category: French or wet kissing and sucking your partner's penis before he ejaculates. Wet kissing is inadvisable because we don't yet know the significance of saliva—or other fluids, such as urine and tears—in the transmission of AIDS. The HTLV-III virus has been found to be present in these fluids.

If you're absolutely unable to avoid "unsafe" sex, limit your number of partners. Another approach would be to engage in sex only with those whose blood tests are consistently negative. Remember, though, that you can only be certain about the AIDS status of your partner within the confines of a monogamous relationship.

A positive test doesn't mean that you have AIDS. It does mean, however, that you have been infected by the HTLV-III virus and are quite possibly infectious to others.

Current estimates are that most people who are infected by the virus do not develop AIDS. Although all the facts about AIDS are not yet known, it appears as if 10 to 20 percent of those infected by the virus will develop full-blown AIDS. An additional 10 to 20 percent will develop milder symptoms.

So even with a positive test, there is no need for gloom. However, there is need for care and responsibility.

It's best to live your life on the assumption that you're going to remain healthy. Statistically, that's the most likely possibility. Worry can only serve to make you sick, as you're already discovering.

If your test is positive, you must assume that you are infectious and will remain so indefinitely. Be sure not to let others come in contact with your body fluids. If you can't live without unsafe sex, have sex only with partners whose blood tests are also positive. You should also know that the health implications of such activity are unknown.

It's important to keep in top shape with regular exercise, a healthy diet, and plenty of sleep. You'll also need to get some counseling. It would be a good idea to contact the Gay Men's Health Crisis AIDS Hotline (212-807-6655) or the San Francisco AIDS Foundation Hotline (415-863-AIDS) for information about counseling groups in your area.

If you can't find a group where you live, then you may want to start one or seek counseling with a knowledgeable professional.

I found my brother (he's 14, I'm 17) and his friends sniffing from a can of engine-starting fluid while masturbating. They were looking at the pictures in a sexy magazine.

I know that it's normal to look at that magazine—I used to do it myself—but what can breathing that stuff do to you? And is it normal to masturbate with your friends?

Should I tell my parents?

You need have no concern about your brother masturbating with his friends. This is so common in adolescence as to be considered normative behavior. Your main concern should be the fact that he is inhaling a toxic substance.

According to Dan Siegel, M.S.W., Director of Broadway Central, a drug counseling center in Ulster County, New York, inhaling engine-starting fluid is much like inhaling glue, paint thinner, or gasoline. The base of engine-starting fluid is ether, and it's mixed with dangerous hydrocarbons, industrial chemicals, and other compounds.

Inhaling engine-starting fluid produces an intoxication much like that of alcohol. At low levels, it's a stimulant. The person becomes less inhibited, sometimes losing all control and becoming unconscious. There's a distortion of vision, judgment is impaired, coordination is reduced, and there's a loss of muscle control.

With occasional use, the effects are usually temporary. But long-term, high-dose use could cause permanent damage to the central nervous system, as well as the liver, kidneys, and bone marrow.

Another important consideration: The product is extremely flammable, so soaked rags could start a fire.

It's estimated that 7 million people under the age of 20 experiment with inhalants. In most instances, they are used in a social setting. Your brother and his friends may have found that the inhalant lowers their sexual inhibitions, possibly enhancing sexual sensations. The inhalant reduces oxygen supply to the brain, which may result in a sense of prolonged orgasm.

Dan Siegel suggests that you talk to your brother. "Tell him you love him and you're concerned about him. Let him know how dangerous the inhalant can be. Find out if inhaling the engine fluid is something that's habit, or if he was just experimenting that once."

If it seems that your brother has a drug problem, urge him to seek counseling. Almost every community now has a hotline for substance abuse. Offer to accompany him to an appointment with a counselor.

If he refuses to seek help on his own, let your parents know about the problem.

Sometimes I get a bad headache when I have an orgasm. What could be causing this?

Headache with orgasm occurs in a small number of people.

If such headaches occur frequently, or if they become increasingly frequent or severe, see your doctor for a complete neurological examination. In rare instances, the headaches could be caused by a blood clot in one of the cerebral blood vessels or by a brain tumor or abscess.

In most cases, orgasmic headaches result from much less serious conditions. High blood pressure may produce a headache at orgasm, particularly if the condition is unsuspected or poorly controlled. The pain may be severe, diffuse, and throbbing, and may persist for many hours.

Other possible causes of headache at orgasm include diabetes and thyroid disease.

A small number of people who suffer from such headaches have been found to have low spinal fluid pressure. They can get quick relief by lying on their backs.

Migraine is sometimes aggravated by orgasm. At the moment of orgasm, some migraine sufferers experience the classic throbbing associated with migraine.

In most cases of orgasmic headaches, however, no physiological cause can be found. It's possible that excessive excitement or effort during intercourse may cause blood pressure to rise above normal levels. Or hormonelike chemicals called prostaglandins, which are produced in both males and females, may be responsible.

Characteristically, headache at orgasm is extremely severe, persistent, and on both sides of the head. Usually, the headaches last an hour or two, but some sufferers have a dull ache of the scalp or neck for several days. For most people, the pain does not occur after every orgasm.

To help prevent orgasm-induced headache, patients are often advised to avoid alcohol before sexual relations. That's because alcohol can further enlarge blood vessels, which may contribute to headache. Drugs such as propranolol and indomethacin have sometimes been used successfully as preventives.

I'm into body-building and running.
Is it true that frequent ejaculation causes fatigue and loss of energy that could affect my athletic capacity?

It's not true. But the myth that ejaculating sperm causes a man to lose his strength has been around for centuries. One expert attributes the myth to the belief that sperm was in some way part of the blood system. If losing sperm was like losing blood, it followed that ejaculation weakened a man.

Because the myth has persisted, athletes have been warned that ejaculation prior to a sports event can mar their performance. That's why many of them avoid intercourse for days or weeks before competing.

There is no justification at all for such unnatural behavior. Provided a man doesn't ejaculate just moments before his event, his performance will not be adversely affected.

I'm going to have a baby by a man who has genital herpes (I'm 26). Will the herpes harm the baby?

Let your obstetrician know that your baby's father has genital herpes. Should you contract herpes from him and have an active infection at the time of delivery, your baby would be at serious risk. That's why you'll most likely be advised to avoid sexual contact with your partner for the rest of your pregnancy.

If you do get herpes, the baby can contract the disease as it passes through the vaginal canal. Without treatment, half of the infants infected with herpes die. Those who survive have permanent neurological or visual damage.

Since your partner is a herpes sufferer, you fall into a high-risk group that also includes women with a history of genital herpes, women with suspicious lesions, and women with oral herpes.

To help ensure the baby's safe delivery, you should be carefully monitored from the thirty-fourth week on. Viral cultures should be taken to see if you have the virus and if it is shedding.

If you are found to have herpes, it should be treated with topical acyclovir ointment, sitz baths, and wet compresses. You should not take oral acyclovir because it has not been proven to be safe for the fetus.

You may be able to deliver vaginally if you have no lesions at the time of delivery and if you have had two consecutive negative viral cultures, one within seven days of delivery. If there is an active infection, a cesarean delivery is recommended. But a cesarean will help protect your baby only if the membranes are intact or have been ruptured for no longer than four hours.

Care is needed not to transmit herpes after delivery as well. As a precaution, both a mother with active herpes infection and her infant are isolated.

If the mother has active lesions, she should keep the infant from any contact with her herpes sores. Before touching her baby, she should wash her hands thoroughly with hot water and soap. Then she should put on gloves, apply topical acyclovir ointment to the lesions, and cover them with a gauze bandage.

Before the mother picks up her infant, she should also change into a clean gown and get out of bed so that the child does not come into contact with linens that may be soiled with secretions from her lesions.

Is it possible for males to be nymphomaniacs? If it is, then I think I might be one.

I feel like going to bed with every attractive girl I see. Is this normal? I'm 17.

Nymphomania is a form of hypersexuality in females. It's a pathologically excessive sexual interest or activity. The same condition in males is called satyriasis.

Satyriasis is characterized by an uncontrollable sexual appetite, an overpowering drive for sexual satisfaction. Sex is the dominant drive in the man's life. The search for sexual gratification is compulsive and indiscriminate.

This is very different from the powerful sexual urges you describe. Your urges are common and normal in adolescents and are in no way related to satyriasis.

One type of satyriasis is called the Don Juan syndrome. The Don Juan unconsciously uses sexual success to overcome feelings of inferiority. He's not interested in his partner of the moment. After proving that he can excite and conquer her, the Don Juan is compelled to move on to the next woman.

The compulsion is fueled by intense narcissistic needs. In some who suffer from the syndrome, a frenzied heterosexual drive is thought to be a defense against unconscious homosexual impulses.

Most often, as in the case of the Don Juan syndrome, satyriasis is psychogenic in origin and is considered a type of compulsive neurosis. It's often treated successfully by psychotherapy. For some men, oral contraceptives taken daily or every other day may reduce sex drive.

Sometimes, though, psychosis may account for satyriasis. During a manic episode, sufferers may exhibit such symptoms as taking multiple, impulsively chosen partners, undressing in public, or masturbating in public. Such behavior usually abates with proper medication.

Satyriasis can also be a symptom of brain dysfunction. Disorders of the temporal lobe occasionally cause compulsive sexual activity in both men and women.

CONFESSIONAL BOX
"I'm no longer a sexaholic"

<u>David, (38)</u>: I just read your response to the question about satyriasis (male addiction to sex). I also have this problem but have found an answer. Several years ago, I read about using the methods of Alcoholics Anonymous in curing one's addiction to sex. Using those methods, I've been making a great deal of progress. I have eliminated the various "triggers" to this behavior, such as pornography, explicit movies, and even masturbation. I had my last affair nearly two years ago and have confined sex to my marriage.

I now run a meeting of Sexaholics Anonymous every week and, from what I understand, there are perhaps some three hundred such groups in the United States. I've learned that part of my problem was an unwillingness to grow up and face my responsibilities as a husband and father. So-called "open marriage" may work for some people, but my experience has been that one member of the couple is usually just going along out of fear of losing the other. I would be happy to share my experience, strength, and hope with any other sexaholics out there.

Since I had an abortion, my period has been lasting for an abnormally long time.
Why should this be?

Mary Anna Friederich, M.D., an obstetrician-gynecologist, recommends that you consult the physician who performed the abortion for a follow-up examination. Your abnormally long periods could be caused, she says, by retention of the products of conception, or a low-grade infection, among other things.

Occasionally, I have what I call "dry orgasms." I have all of the normal orgasmic reactions, but there is no discharge of semen.
I'm 27 and diabetic. I usually discharge a normal amount of semen, and I have intercourse two to four times a week. Also, at times my semen has a yellowish tinge to it.

You should bring the "dry orgasms" to the attention of your physician.

You might be experiencing a condition called retrograde ejaculation, in which semen is ejaculated into the bladder rather than out through the urethra. A common cause of retrograde ejaculation is nerve damage due to diabetes.

Sexual performance is not affected by retrograde ejaculation. But don't count on it as birth control. As you've seen, the condition may occur unpredictably.

Why the yellowish tint to your semen? Terrence Malloy, M.D., a urologist, says that discolored or blood-tinged semen may sometimes be caused by an infection in the genitourinary system. Have this checked out too.

I'm worried that my penis is too small to have sexual intercourse (I'm only 15, so maybe it will grow more).
How long is the average vagina?

In all likelihood your penis, whatever its length when erect, is perfectly adequate for sexual intercourse. And it's also possible that it will grow more in the next few years.

The size of the vagina varies from woman to woman. Generally, it is three to five inches long. During intercourse, the vagina is capable of expanding to accommodate penises of varying sizes.

Rather than being like a hollow tube, the vagina is like a balloon or a sock that can stretch. Like a collapsed balloon or a sock, the walls of the vagina touch each other. When the penis penetrates, the walls are parted, and the vagina expands. That's why the vagina can adjust to any size penis during sexual intercourse.

I've been too embarrassed to seek help for my problem.

When I develop an erection, my penis does not become erect enough to point straight out. It stiffens, goes into a sharp bend in the center, and continues into a curve.

When not erect, it looks slightly discolored and out of shape. If this can be treated, please tell me where to look for help. I'm 17, and I've had this problem for as long as I can remember. Needless to say, I am a virgin.

It's very likely that your problem can be corrected. You need to consult a board-certified urologist for advice. There's no need to be embarrassed. Such doctors are used to problems of this kind.

From your description, you may be suffering from a fairly common condition called chordee. In chordee, there is a fibrous band of tissue around the urethra, the tube through which urine passes. The presence of the tissue results in an abnormal curvature of the penis. With erection, the curvature becomes greatly accentuated.

If uncorrected, chordee may interfere with sexual intercourse. A pronounced curvature can make vaginal penetration difficult or impossible. A downward curve may make intercourse uncomfortable for one or both partners.

Surgical correction—in which the fibrous band is removed—usually gives good results.

Chordee is often associated with urethral problems, although it may also occur alone. When, in addition to chordee, an infant has an abnormality such as the urethral opening on the underside of the penis, chordee is likely to become apparent at birth. Surgery is usually performed when the boy is between two and four years old.

Since parents rarely see a child's erection, your parents may be unaware that you have a serious problem requiring medical attention. It's wise to confide in them and ask for their help in locating the best medical care.

If you're too embarrassed to tell them the whole story, simply say that you suspect you have a urinary problem, and you'd like to consult a urologist.

I've been having inflammation of the prostate— congestive prostatitis—on and off for the past three years (I'm 37 now).

Why is this happening? Is there any way to cure it once and for all?

Congestive prostatitis usually occurs in men who are experiencing a marked decrease in sexual activity.

The prostate gland is a small organ located at the base of a man's urethra. At ejaculation, the muscles in the prostate squeeze the embedded glands to discharge a fluid that helps to transport and nourish sperm.

During sexual arousal, fluid in the prostate increases. If a man repeatedly becomes sexually aroused without ejaculating, the prostate may become congested with fluid, causing inflammation. This may also happen if a man's frequency of ejaculation suddenly declines. Common symptoms of prostatitis include pain on ejaculation, an aching in the groin and testicles, pain on urination, and frequent urination.

Your congestive prostatitis may be due to changes in sexual activity. If a marriage is troubled, and sexual activity decreases, the prostate can become inflamed. Or your preoccupation with work-related stresses may have led to a reduction in sexual frequency.

Sometimes, congestive prostatitis affects men who have "feast or famine" sex lives. A traveling salesman, for example, may have frequent sex at home but abstain while on the road. His prostate, used to regular sexual activity, becomes congested and inflamed. Much more rarely, congestive prostatitis results from an unusual increase in the number of ejaculations over a short period of time.

A diet heavy in spicy foods, chocolate, alcohol, and caffeine may aggravate prostatitis. So can sitting for long periods of time, riding on bumpy roads, and emotional stress.

The best way to control prostatitis is to establish a regular pattern of ejaculation. Frequency is not as important as regularity. Whether ejaculation occurs once a day, once a week, or once a month, you and your partner should try to maintain the pattern that's most comfortable for you.

To help relieve symptoms, your doctor may recommend sitz baths twice a day in very hot water. You should also avoid alcohol, spicy foods, coffee, tea, cola, and nuts. Cashew nuts, in particular, increase prostatic fluids.

If the prostatitis continues to recur despite these measures, have your doctor check to see whether the condition is caused by a bacterial infection. Chronic bacterial prostatitis is a common condition, particularly in men over 50. It can usually be cured by prolonged treatment with the drug trimethoprim (sold as Bactrim or Septra).

I like to have anal sex with my girlfriend. But I'm afraid when I ejaculate that the semen could cause an infection in her rectum.

We both enjoy this act very much. Is there any possibility of infection for either of us?

There is nothing in the semen itself that poses a threat of infection. But if you have a venereal disease, it can be transmitted through anal sex and cause infection in the rectum.

The most common anorectal venereal disease is gonorrhea. Rectal symptoms, if there are any, range from mild itching or burning to severe discomfort.

Many other venereal infections—including syphilis, anal warts, chancroid, granuloma inguinale, AIDS, and herpes—can also be transmitted by you through anal intercourse.

The converse is also true: If your girlfriend has a venereal infection in her rectum, it can be passed to you. There's also the possibility that you could acquire urethritis (inflammation of the urinary passage) from bacteria that normally inhabit the rectum, such as *Escherichia coli*. But this has not been well documented.

To be on the safe side, wear a condom for anal intercourse.

I've heard that oral sex can sometimes be fatal if a woman is pregnant. Is this true?

Yes, cunnilingus can sometimes be fatal if a woman is pregnant.

When air is blown into the vagina of a pregnant woman, a lethal air embolism can result. Bubbles of air can pass through the open cervix into the dilated uterine blood vessels surrounding the fetus. Once in the bloodstream, an air bubble may be carried to the lungs or the brain. If the bubble blocks circulation, coma and death may result.

Thus, the practice of blowing air into the vagina should be avoided if there is the least possibility that a woman could be pregnant.

I have only read about the rape of women. Can men get raped?

Good question.

Rape of males is increasingly recognized as an important social concern—and it's much more prevalent than most people think.

"Sexual victimization of males is much more common than most people recognize, but we don't talk about it in our culture," says Dr. Margaret T. McHugh, director of the child-abuse team at Bellevue Hospital in New York City.

Male rape victims are much less likely than women to report the event. Dr. McHugh estimates that 20 percent of sex abuse cases involve males, "but the statistics are meager and confusing because so few cases are ever reported."

Males who have been raped fear they will be branded as weak or unmasculine if they reveal the assault. They are even more likely than women to feel responsible for the event. They think they should have fought to the death. They also fear being stigmatized as homosexual if they report a sexual attack by another male.

Professionals agree that, like the rape of females, male rape is not primarily a sexual act. Rather, "sex becomes the way in which the offender compensates for feelings of inadequacy and vulnerability and discharges pent-up anger and resentment." So says Dr. A. Nicholas Groth, author of *Men Who Rape* and director of the Sex Offender Program at the Connecticut Correctional Institution.

Men who rape other men are not necessarily homosexual. Indeed, "the sexual attack may be an expression of contempt on the part of the offender toward someone he perceives as homosexual," says Dr. Groth. Or it may be a way of asserting his superiority and manhood.

Besides rape in prison settings, sexual assaults on males also typically occur during break-ins or robberies. In adolescent gang attacks, rape may be used as the ultimate form of humiliation.

Men who have been raped display the same symptoms as female rape victims. They are commonly fearful, anxious, helpless, and angry for weeks or months after the rape. They may experience nightmares, loss of appetite, sexual disturbances, and psychosomatic ailments.

In a study that compared male and female rape victims, the male victims as a group "sustained more physical trauma," reports Dr. Arthur Kaufman of the University of New Mexico in Albuquerque. Men were also more likely to have been the victims of multiple assaults by multiple assailants, and to have been held captive longer.

I don't sleep with any men besides my husband, who had a vasectomy two years ago. I suspect that I may be pregnant. Is this possible?

Yes, there is a small possibility of your becoming pregnant more than a year after your husband had a vasectomy.

In rare cases—estimated at 6 in every 1,000 vasectomies—the severed ends of the vas deferens grow together and fertility is accidentally restored. That's why many physicians recommend semen analysis every other year for six years after the vasectomy.

A Peeping Tom, or voyeur, obtains sexual gratification by observing others undressing or engaging in sexual activity. To some extent, almost all of us are voyeurs. It's almost universal to be interested in what other people's bodies look like.

During childhood and adolescence, occasional peeping is fairly common, especially among males. Often, adolescent boys will go around in small groups and peep into windows in the hope of seeing a female undressing.

However, when an adult has a compulsion to peep, and it's his only means of sexual gratification, he is considered to be in need of therapy. Voyeurs will rarely observe women they know. It's also rare for them to approach the women they view, so they are hardly ever dangerous.

"There is little evidence that this disturbance is serious, at least in the sense of being dangerous to others," observes Donal E. MacNamara, Distinguished Professor of Criminology at John Jay College of Criminal Justice in New York.

Most Peeping Toms will peer hopefully through windows at night. Some bore peepholes through walls of bathrooms or dressing rooms in public places. Sometimes, voyeurs masturbate while peeping.

While voyeurs pose little or no danger, Dr. MacNamara points out that it's not easy to distinguish them from burglars. So people are justifiably concerned when they find a man lurking about at night.

The typical adult voyeur is a man who is very shy with women. He is generally passive, socially inhibited, and fearful. Usually, he has little sexual experience and strong feelings of inferiority. He may experience sexual difficulties in heterosexual relationships.

According to psychoanalytic theory, as a child the voyeur may have witnessed his parents having intercourse. His adult voyeurism may be a defense mechanism against an act he perceives as threatening. By merely looking, the voyeur guards against personal failure in sexual activity. At the same time, he can enjoy a feeling of superiority over those he is secretly observing.

Voyeurs can often benefit from individual or group therapy.

I like to be very clean when I have oral sex. Is it a good idea for me to douche beforehand?

It's unwise and unnecessary to douche before oral sex. Medical authorities note that the vagina is a self-cleaning organ. In fact, douching can remove the helpful bacteria that usually protect the vagina.

It's understandable that you want to feel clean when having oral sex. But simply washing your genitals is all you need to do.

My wife had a hysterectomy a few months ago. She's 40 years old. Her uterus was removed, but not her ovaries.

Now she seems to have no interest in sex. Is this because of the operation? Can anything be done?

The removal of a woman's uterus need not interfere with her sex drive or her ability to enjoy sex.

The vagina and clitoris, which respond to sexual stimuli, are not affected by removal of the uterus. If the ovaries are not removed, the woman's hormone production remains unaffected.

The only physiological effect of the surgery is that the woman will no longer be able to bear children. As a matter of fact, many women report increased pleasure in sex after hysterectomy. One of the reasons may be that they no longer have to worry about pregnancy.

However, if a woman's enjoyment of orgasm before surgery was caused by contractions of the uterus, she may find that her orgasm after hysterectomy is somewhat changed.

Like your wife, some women experience decreased sexual interest after hysterectomy. Sometimes the decrease is connected to the reasons for the surgery. For example, a woman whose uterus was removed because of cancer of the cervix may be anxious and depressed. If a woman had a hysterectomy because of severe fibroid tumors that caused heavy menstrual bleeding, she may be fatigued due to iron deficiency.

Some women have less interest in sex after a hysterectomy because the loss of the uterus makes them feel less feminine and less sexually desirable. Some think their husbands may no longer see them as fully female, since their reproductive years are over. Thus, they withdraw from sexual contact.

Since there are so many reasons why a woman may lose interest in sex after a hysterectomy, your best course is to consult a physician. If physical causes are ruled out, you may want to ask for a referral to a sex therapist or psychological counselor.

I know that women can do vaginal exercises to improve their enjoyment of intercourse.
Are there any comparable exercises for men?

Yes.

The penis has an elaborate network of muscles, similar to the musculature of the vaginal area. Strengthening these muscles may produce benefits for men that are similar to those women get from doing vaginal Kegel exercises.

After doing the male Kegels, men have reported stronger and more pleasurable orgasms, better ejaculatory control, and increased pelvic sensation during sexual arousal.

In his book *Male Sexuality*, Dr. Bernie Zilbergeld recommends the following program for strengthening the penile muscles:

• Locate the muscles by stopping the flow of urine several times. The muscles you squeeze to do this are the ones you will be working on.

• To begin the exercise program, squeeze and relax these muscles fifteen times in quick succession. Do this twice a day. The maneuvers are called ''short Kegels.''

• Gradually increase the number until you can easily do sixty short Kegels at a time, twice daily.

• Next, start practicing ''long Kegels,'' holding each contraction for a count of three.

• Combine the short and long Kegels in each daily exercise routine, doing two sets of sixty shorts and sixty longs per day.

You can expect to notice results in about a month or so.

As the mother of two teenage girls, I'd like to be able to advise them—when the time comes—about birth control.
What contraceptive would you recommend?

There is no one contraceptive that is appropriate for all teenage girls. The choice of a method depends on how often a girl has intercourse and how likely she is to plan ahead for it.

Most teenagers have intercourse infrequently. Researchers Melvin Zelnik and John Kantner of Johns Hopkins University found that fully 38 percent of the sexually active girls they questioned had no intercourse at all during the month before the interview. Thirty percent had intercourse once or twice in that month, and 18 percent had intercourse three to five times.

Only 14 percent of the girls had intercourse six or more times during the month. In contrast, married couples in which the wife is under 25 average intercourse nine times a month.

In addition to having intercourse sporadically, many teenagers don't reliably prepare for it. Often, they feel guilty about making such plans and, therefore, don't.

Since teenagers tend to avoid long-range planning, a good method of contraception is contraceptive foam plus a condom. The materials are inexpensive, widely available, and require no prescription. The method is also extremely effective and has no side effects. Furthermore, the condom provides protection against many venereal diseases that are epidemic among teenagers.

Using a condom plus foam also has the advantage of having the boy share responsibility for avoiding pregnancy. When both parties face that possibility, it's more likely that contraception will be used.

A diaphragm with a condom is even more effective than the condom with foam, and it's equally free of side effects. It also protects against many sexually transmitted diseases and is a shared responsibility. But the diaphragm requires a physician's prescription and some instruction in its proper use. It's suited mainly to mature young women who can be counted on to use it properly.

The diaphragm is also more appropriate for girls who have intercourse frequently. Otherwise, it may not be worth the investment in money and time. Keep in mind that the diaphragm is a very poor choice for an immature girl.

Although an intrauterine device (IUD) has been recommended by some physicians for girls who are sexually active, some of the latest evidence suggests that it may not be a good choice.

In "Good News about Contraceptives and PID" by Andrew M. Kaunitz, M.D., and David A. Grimes, M.D. (*Contemporary Ob/ Gyn*, March 1986), the authors found that women with IUDs who have not had children and who have a history of multiple sexual partners run a higher risk for developing tubal infertility due to pelvic inflammatory disease (PID). Young women should consider this risk and discuss it thoroughly with a doctor before deciding about an IUD.

Oral contraceptives (birth control pills) are a poor choice for many teenage girls. Usually, teenagers do not have intercourse frequently enough to justify the risk of side effects.

What's more, many teenage girls are not disciplined enough to remember to take the Pill every day. And skipping even one day may make them vulnerable to pregnancy.

As for old-fashioned "methods" such as withdrawal and douching—they are very ineffective methods of birth control and should not be recommended to your daughters.

I'm the mother of an eight-year-old boy who seems to have an erotic attraction to enemas.

He asks me to give him enemas, even though I stopped some time ago because of what seemed to be an unhealthy sexual interest. But I find him still trying anal masturbation when he thinks I won't find out. He uses douche syringes and other enemalike objects.

Does he need counseling to be cured?

Dr. Richard V. Lee, an internist with a special interest in the health problems of young people, thinks it would be a good idea for both you and your son to seek some counseling. It's hard to say that your son has an ''unhealthy sexual interest.'' All children have sexual interests. But his choice of expressing that interest is unusual.

A careful professional evaluation of your son may help you understand what's motivating his behavior. Many unusual sexual styles derive from conflicts and distress in other areas of life. When these are addressed and resolved, the unusual sexual style may diminish.

I'm a 42-year-old man with diabetes, and I'm having a worsening problem with impotence. My doctor says I should consider a penile prosthesis before long.

How do they work? What kinds are there?

Penile prostheses are an option for men who are unable to have erections.

Most candidates for these surgical implants are men like you who are impotent because of irreversible organic causes. Besides diabetes, common causes of organic impotence include neurologic disorders, vascular conditions, and surgical procedures that interfere with erection. Sometimes, penile implants may be recommended for men who have psychogenic impotence when psychotherapy and sex therapy have failed.

There are two general types of penile prosthesis: the semi-rigid rod prosthesis and the inflatable prosthesis.

Semi-rigid rod devices consist of two silicone rods that are surgically implanted in the penis. Among the semi-rigid devices are the Small-Carrion prosthesis and the Jonas-Silver prosthesis. The

Finney (flexi-rod) prosthesis is a similar device with a hinged pair of rods that permits the penis to hang down when the man stands.

Semi-rigid devices have enough length and girth to simulate a normal erection and to allow vaginal penetration. These devices present few mechanical problems; they require relatively simple and inexpensive surgery.

They do, however, result in a constant state of partial erection. This must be concealed by using an athletic supporter and wearing brief-style underwear. Some men may find this permanent erection an embarrassment. It may be particularly troublesome for younger men who engage in competitive sports.

The inflatable penile prosthesis (also called the Scott-Bradley) is designed to mimic the normal erectile response. It distends the penis when erection is desired and is deflated after ejaculation so that the penis becomes flaccid.

The device has three components: twin cylinders, a reservoir, and a pump with a release valve. The hollow, twin cylinders are implanted in the penis. A reservoir containing radiopaque fluid (which can be seen on X-ray) is implanted in the abdomen. The pump is implanted in one scrotal sac. The cylinders, reservoir, and pump are connected by silicone tubes.

The entire device is implanted through an incision in the scrotum just below the base of the penis. When the man desires an erection, he squeezes the valve in the scrotum. Squeezing makes the solution flow from the reservoir in the abdomen and causes an erection. A one-way valve keeps the solution in the cylinder.

When the man wants the erection to subside, he compresses the release valve, and the fluid returns to the reservoir.

Many men find the inflatable device more psychologically satisfying. It more closely resembles normal functioning than the semi-rigid devices.

When not erect, the penis has a normal appearance. On erection, the inflatable device has greater length and girth than the semi-rigid rod prostheses. But the erection with the inflatable implant will not look exactly as it did before the onset of impotence, largely because the head of the penis does not swell.

The inflatable penile prosthesis has been inserted in more than twelve thousand men. It requires more complex surgery and is much more expensive than the semi-rigid rod devices.

It's also more subject to mechanical failure than the simpler rod devices. There may, for example, be kinks in the tubing or a malfunction of the hydraulic system. Such problems can almost always be easily corrected, but they require more surgery.

The inflatable device may not be suited to men with severe physical handicaps who might find it hard to manage.

Sexual activity can usually begin four to eight weeks after the surgical insertion of either prosthesis. Normal orgasm and ejaculation can occur with both types of devices—unless the man's medical condition makes this impossible. If the man's sperm production has not been affected by the cause of the impotence, he can also father a child.

I am a gay man who likes giving oral sex to other men. Gay men are now being advised not to swallow ejaculate. But I find this one of the pleasures of oral sex. What do you think?

Fred Westendarp, M.D., says there's little doubt that sexually transmitted diseases such as herpes, hepatitis, gonorrhea, and syphilis are frequently spread by oral sex.

It hasn't been proved that spitting out the semen or gargling after sex have any effect on preventing the spread of VD. But even if these measures only reduce the risk a little bit, wouldn't they be worth the inconvenience? You have to decide.

AIDS is a different matter. Here the high risk of swallowing semen is known. It's best for gay men to avoid oral-genital sex. But for those who continue to engage in it, the Bay Area Physicians for Human Rights Scientific Affairs Committee and the San Francisco AIDS Foundation offer the following guidelines for reducing the risk of AIDS through oral-genital sex:

• Sucking your partner's genitals presents minimal risk if his penis and your mouth have no cuts or sores. But remember that semen can contain germs or other material foreign to your body. To protect yourself, either don't have your partner ejaculate in your mouth, or don't swallow the ejaculate.

• If there is uncertainty about the recent whereabouts of your partner's penis, it can't hurt to request that he take a shower.

• If you're being sucked, and there are no cuts or abrasions on your penis to be infected by mouth germs, you probably need take no extra precautions except to avoid trauma done by the teeth. But remember to protect your partner's health by not ejaculating in his mouth.

You can get copies of *Guidelines for AIDS Risk Reduction* from the San Francisco AIDS Foundation, 54 Tenth Street, San Francisco, CA 94103-0960.

My neighbor shocked me by confiding that she has never had an orgasm, but she pretends to have one every time. I've always thought she and her husband were a perfect couple.

Is faking orgasm a common practice among women?

It's not uncommon. According to *The Hite Report: A Study of Female Sexuality*, more than half of the over 1,600 women interviewed said that they currently faked orgasm or did at one time. About two-thirds of the women in the study said that they need additional stimulation beyond intercourse to reach orgasm.

The research of Masters and Johnson has shown that female orgasm is clitorally triggered by direct and indirect stimulation. Many couples are not aware of the role of the clitoris in sexual pleasuring, nor do they know techniques for its stimulation.

A woman may fake orgasm because she fears hurting her partner's ego or being called "frigid." When she pretends, she makes no demands on his performance by disregarding her own needs. A woman may also fake orgasm to terminate the sex act if she sees intercourse as a "duty" rather than a shared pleasure.

If this situation exists over a long period of time, the woman may have difficulty in telling her partner about her dissatisfaction. As she becomes resentful of her situation, she may avoid sexual contact altogether.

There may be physical discomfort for a woman as well as psychological consequences from faking orgasm. When no orgasm takes place following sexual excitement, the blood that has pooled in the pelvic area takes longer to return to the system. As a result, a woman may experience pelvic pain and a feeling of restlessness.

To achieve their orgasmic potential, some women need only learn more about their bodies in order to help their partners help them achieve orgasm. Other couples find that joint counseling is needed to resolve the problem.

Ever since I was a teenager (I'm 38 now), I've been attracted to young boys.

I'm married, and I have a very pleasant sexual relationship with my wife. I had good sexual relationships with women before I married.

However, I am tremendously excited by boys who are in early puberty through the age of 17. Although I would never seduce a boy or pay a teen hustler, I have had sex with several boys who openly wanted to have sex with me.

Am I contributing to psychological problems for those boys?

Richard V. Lee, M.D., a pediatric internist, points out that teenage boys have sex for reasons that are as diverse and individual as those of adults.

Even though your adolescent sex partners appear to desire sexual relations with you, participating in homosexual relations with an older, married man may lead to serious psychological confusion. Your activities may also have an adverse effect on your wife, and you risk infecting her with a sexually transmitted disease.

As for you, the fact that you ask your question indicates that you may be having self-doubts. Perhaps you're worried that should your inclinations become known, your career and your marriage would be jeopardized.

You should be seeing a therapist about this problem. While you feel in control of your sexual urges now, your question implies a serious, perhaps latent, doubt about the future.

CONFESSIONAL BOX:

"I need a teenage boy"

<u>Bob, (37)</u>: I happen to be a pederast, someone who likes young males in their early teens. Okay, so I should seek help. I have, and it didn't work. The result is a life of repression. I cannot morally justify getting involved with a minor. Yet, I have desires. Talk about being an outcast. Heck, I would be happy just being gay! Feeling like a pervert is living hell, but living out the fantasies would be wrong and dangerous. Life's a bitch and then you die, right?

I've read that a particular type of IUD was recalled by the manufacturer. Since I've been wearing an IUD for many years, I'm concerned. What's the story?

You've most likely read that the Dalkon Shield intrauterine birth control device, sold between 1971 and 1974, was recently recalled by its manufacturer, A. H. Robins Co.

Sales were suspended when the Shield was reportedly linked with thousands of cases of pelvic inflammatory disease, some leading to sterility, miscarriage, and death. A. H. Robins urges any woman who is wearing the Dalkon Shield to have it removed, at the manufacturer's expense.

Thousands of women in this country are believed to be using the Shield, unaware of its potential dangers. If you suspect you may be among them, see your gynecologist.

For information about the recall, call the manufacturer's toll-free number: (800) 247-7220. In Virginia, call (804) 257-2015 collect. Information about the Dalkon Shield can also be obtained from the National Women's Health Network, (202) 543-9222.

I recently found out that I'm a carrier of hepatitis B. With whom can I have sex without passing on the disease?

Someone told me that if a partner had antibodies against hepatitis B, she would be safe.

Nicholas J. Fiumara, M.D., a venereologist, confirms that people who have developed hepatitis B antibodies are safe sex partners for carriers. Make sure that a potential partner actually has the antibodies, though. You're correct in thinking that hepatitis B—also called viral or serum hepatitis—is often transmitted through sexual activity.

Dr. Fiumara advises that the spouse or sex partner of a carrier have a blood test. The test has two parts. The first part reveals if the person has the disease now. The second part indicates if the person had the disease at one time and now has antibodies that make him or her safe from transmission.

I'm the mother of a 13-year-old boy. Recently, I was stunned to open a door and find him having sex with a girl the same age. I stopped the activity. But I'm confused. What should I have done?

Sex educator Judy Henkel of Planned Parenthood says that you needn't feel uncomfortable about stopping sexual intercourse between youngsters. Remember that it's your home, and you are entitled to set the limits of behavior. If you object to young teenagers having intercourse in your home, tell them so.

A good way to approach a situation like this is to say to the couple: "I would like you to stop now, get dressed, and then come out and talk to me." The youngsters must be given time to respond. They'll be embarrassed and have to get themselves together.

When they emerge, you might comment, "I'm not comfortable with this behavior. It offends me in my home."

Present the "no sex" rule in the context of other family rules. It's unlikely that your son is allowed to shoot baskets in the living room, for example. In the same way, you can point out that having sex in your home is inappropriate.

Much of your discomfort may have to do with your son's young age. You might not have the same feelings if he were 19 or 22 and was discreet about his behavior. Or, you might. It's up to you to set the standards. Whatever your position, state it to your child. Letting him know your sexual values will aid him in formulating his own.

You should also talk to your son about the realities of having sexual intercourse at such a young age. You can tell him: "It's best to give yourself time to grow up before engaging in sexual intercourse. At your age, people are easily hurt and easily exploited. They may not realize how sexual intercourse will affect them emotionally and physically." Point out that for the time being, hugging and kissing are much more appropriate ways of expressing affection.

Make sure your son is aware that he could impregnate a girl. Many young teenagers do not really accept this possibility as a consequence of sexual intercourse. Some may believe that they are too young to cause pregnancy. Others refuse to take the possibility seriously. With such attitudes, they're unlikely to use contraception reliably—if at all.

Also talk to your son about sexually transmitted disease. Young teenagers are likely to be uninformed about the many venereal diseases that are epidemic.

At the same time as you express your views, it's important for you to be realistic in your expectations of your child. You can set standards in your own home, but you may be unable to enforce them elsewhere.

Forbidding a child to have intercourse at all is often ineffective.

The best you can do is to give your son good advice and practical information, then hope that he will make a responsible decision on his own. If it seems likely that your son will continue to have intercourse outside the home, make a point of discussing birth control choices.

CONFESSIONAL BOX
"Parents should wake up"

<u>Andrew, (17)</u>: The worst thing parents can do when their child is maturing is to ignore the fact and continue to treat him as a kid. That's what my parents did, and since I was maturing rather early anyway, I had to face the problems of adolescence alone. My social life was pitiful; I was the class nerd. After a couple of years, they're finally beginning to realize that I'm growing up. I wish someone would tell the parents of other teenagers to at least acknowledge that their child is maturing—and react to it in some way.

I have been masturbating heavily for the past few days (eight to ten times a day). Today, while attempting to masturbate, I ejaculated blood! What's the matter with me? (I'm 16 years old.)

Here are reassuring words from Randall Rissman, M.D., a family physician: "Blood in the ejaculate is not usually associated with a serious illness. Chances are that in this case the blood comes from the rupture of a small blood vessel, and that it was caused by frequent masturbation."

Dr. Rissman advises that you refrain from masturbating for a few days—or at least do so less often—so as to give the blood vessel a chance to heal.

If you continue to find blood in your semen, consult a physician. In some instances, it is a sign of infection in the genitourinary tract.

The first sign of syphilis is usually a sore called a chancre. It appears from ten to ninety days after exposure—the average is twenty-one days.

The primary chancre appears on the penis, in the vaginal area, or in or around the mouth or anus—wherever the syphilis germ first entered the body. If a syphilis sore is on the mouth, the disease may have been transmitted through kissing. The chancre may look like a pimple, a blister, or an open sore. Rarely does it hurt or itch.

The chancre is one of the most contagious forms of syphilis. It contains millions of spirochetes, the syphilis bacteria. These germs can be passed on to any person whose mucous membrane tissues come in contact with the sore. An untreated person can remain infectious for two years or more.

The primary chancre may not appear at all in some people. Or the chancre may be so small that it's overlooked. In women, it may be hidden inside the vulva or vagina. The chancre can easily be ignored or mistaken for something else. Some people, not realizing that they have a syphilis sore, attempt self-treatment with salves or ointments. With or without treatment, though, the chancre will heal spontaneously within a few weeks.

But this does not mean that the disease is cured. On the contrary, it has gone underground, where it continues to do severe damage to the body.

If you suspect you have been exposed to syphilis, go to your doctor or to a VD clinic as soon as possible. Syphilis can be detected through a simple, inexpensive blood test. It's usually treated with large doses of penicillin.

If it remains untreated, syphilis can be very dangerous. You may feel perfectly well for many years, while the infection attacks your heart and central nervous system. In its late stages, syphilis can cause blindness, syphilitic insanity, and death.

Chances are a tampon can be inserted into the rectum with no ill
effects, even though putting any object into the rectum poses some
danger.

When a tampon is inserted into the vagina during menstruation,
the menstrual blood helps ease the process by providing lubrication.
In contrast, the rectum produces little natural lubrication. That's
why a tampon applicator in the rectum can injure delicate tissues.

There's also the possibility that an object placed in the rectum
may move out of reach. In the vagina, a tampon cannot get lost,
because the opening to the uterus is much too narrow. But an object
pushed into the rectum too deeply can get lost in the sigmoid colon,
which joins the rectum at a right angle. If that happens, a physician
may have to retrieve the object, possibly with special instruments
and anesthesia.

Besides the medical dangers, you might be concerned about the
psychological dangers of joining such a club. The fact that you
asked this question indicates that you have reservations about performing the rite—and you're ambivalent about joining the club as
well.

To help sort out your feelings, ask yourself what the rite means.
Is this an all-male club in which antifemale feeling is shown by
making the tampon a special object of disgust? Or is the purpose
of the rite to humiliate and demean you? Are you being forced to
show your absolute willingness to follow orders, whatever those
orders might be? If the answer to any of these questions is yes, you
should ask yourself if this is the sort of group you want to be a
part of.

■ *I just got an IUD. Someone told me that this increases my chances of having an ectopic pregnancy.*
What is an ectopic pregnancy? How do you know when you have it? What can you do about it?

An ectopic pregnancy is one that occurs in an abnormal place.

Normally, the fertilized embryo implants itself in the uterus. But in a small number of women, the embryo becomes implanted outside the uterus, in a place where it cannot survive. That place may be the fallopian tubes (the most common place for an ectopic pregnancy), the cervix, an ovary, or the abdominal or pelvic cavity.

Wearing an IUD does put you at greater risk of ectopic pregnancy. One out of twenty pregnancies that occur with an IUD in place is ectopic.

In about half the cases of tubal pregnancy, implantation is caused by a previous infection in the fallopian tube. Scar tissue may trap the fertilized egg instead of allowing it to pass to the uterus.

The first signs of tubal pregnancy—spotting and cramping pain in the abdomen—mimic the signs of a threatened miscarriage. They usually begin shortly after the woman has missed her first menstrual period. When a woman experiences such symptoms, she should consult a physician at once.

If a tubal pregnancy continues undiagnosed beyond eight weeks, the woman may suddenly feel severe pain in her lower abdomen. She may faint. These symptoms usually indicate that the fallopian tube has ruptured and is hemorrhaging into the abdominal cavity.

Whether the tubal pregnancy is diagnosed before or after rupture, the treatment is surgical removal of the tube. The embryo implanted in it is removed as well. Sometimes the ovary on the involved side must also be removed.

■ *My girlfriend and I discovered that if she applies pressure just behind my scrotum at the point of orgasm, I won't ejaculate. But I'm worried that the prolonged use of this technique could have an adverse effect.*

Terrence Malloy, M.D., a urologist, advises that the practice should not be continued.

Pressure behind the scrotum causes occlusion (closing) of the urethra, the tube that delivers urine from the bladder and semen from the prostate and seminal vesicles.

This technique can cause back pressure on the urethra and bladder with potential irritation to the bladder or even the prostate. Prostatitis or urethritis can result.

I'm 21 and my girlfriend is 20. A year before we met, she was raped.

Early in our relationship, she didn't object to being touched. But now she says she feels very dirty when we touch. I feel somewhat as she does because I was sexually molested as a child.

We both have an interest in sex, but we're unable to have real physical intimacy because of emotional problems. What can we do?

It's not unusual for women who have been raped to experience problems with physical intimacy for months or even years after the event. Some develop an aversion to sex or to activities that are associated with the rape.

If your girlfriend did not seek counseling after the rape, she had no opportunity to talk with other women who had the same experience. Nor did she have the chance to vent such feelings as anger, helplessness, and guilt, which are common after rape. When such feelings are unresolved, they can contribute to the difficulties you describe.

Your girlfriend could be helped by contacting a local rape crisis center. They're found in every large city and in most smaller cities and towns. A center's most important function is to offer counseling to women after rape. Although most counseling takes place immediately after the rape, it can still be beneficial years afterward. Look in the yellow pages for such a center, or consult your local hospital or police department.

Your girlfriend could also visit a local mental health clinic or a therapist who specializes in helping rape victims. You should seek help too, since your own unfortunate childhood experiences are contributing to the problem.

Genital warts (condylomata acuminata) are caused by a virus similar to the virus that causes warts elsewhere on the body. The warts are generally painless, but may sometimes be itchy. They usually appear as soft, moist, pinkish swellings that may grow together to become large cauliflowerlike structures.

Genital warts can be transmitted sexually. In a study of ninety-seven patients who had intercourse with partners known to be infected with venereal warts, sixty-two became infected. The average incubation period was about three months. However, medical texts usually say that genital warts are "generally" transmitted by sexual activity—leaving open the possibility of nonsexual transmission.

Your partner may have the disease in spite of the fact that he has no visible warts. It's possible that he has them hidden in his urethra or rectum and has transmitted them to you. He needs a thorough examination to determine whether he has warts, too.

In the meantime, refrain from sexual activity until all warts are treated. Follow-up examinations are recommended, since venereal warts tend to recur after treatment.

Very likely, according to several studies of barbiturates (such as Seconal) and other sleep-inducing medications.

When a person first starts taking such drugs, he or she may find they reduce sexual inhibitions, thus enhancing sexual enjoyment. But chronic use of sleeping pills can decrease sexual desire. Often, there's also difficulty in attaining orgasm. Men may become impotent. Women may experience menstrual abnormalities.

I'm an 18-year-old male who is still a virgin. Sometimes I feel ashamed, but I haven't found the right girl yet. Am I normal?
What's the correct age to lose your virginity?

There's no need to be ashamed. And there's no correct age for beginning sexual intercourse. You're best off following your own sexual timetable. Do what seems right for you.

It's perfectly normal to be a virgin at your age. A conspicuous group of sexually active teenagers may make it appear as if "everyone is doing it." That is simply not true. While sexual activity among teenagers has indeed increased in recent years, a very large number of teenagers refrain from intercourse, especially in the middle class. Somewhat more than half of all teenagers have intercourse before they're 18. But bear in mind that nearly half do not. So you are by no means alone in your virginity.

Much of what is proclaimed as sexual experience is actually hot air. Reports Dr. James Toolan, a psychiatric consultant at Bennington College in Vermont: "Youngsters of both sexes sometimes brag about their sexual experiences even though they may be virgins."

Losing virginity does not mean that a person is grown up. A penis in a vagina does not automatically confer adulthood upon either partner. A person can surrender his or her virginity and still not have experienced meaningful, caring sex. The compulsion to lose one's virginity can lead to what sociologist Lester Kirkendall describes as the "exploitative, self-centered, injurious use of sex as contrasted to a meaningful and dignified use."

Follow your own instincts and wait until you are ready.

If I suck on my wife's breasts during lovemaking, is it possible that she'll start giving breast milk again?

Roy E. Brown, M.D., associate professor of community medicine at Mount Sinai School of Medicine in New York City, points out that "the infant's suckling is generally very different from breast stimulation . . . during lovemaking."

"To begin with," he says, "the infant must grasp the nipple and areola and both suckle and compress this part repeatedly over the usual nursing period of 20 minutes up to six or more times daily, in order for the non-lactating woman to be induced to lactate.''

However, comments Dr. Brown, it is possible to induce milk flow in a woman who has not recently—or even who has never— had a child. "Relactation has been accomplished mainly by strong and vigorous suckling over a period of time, determined by the state of health of the woman, her age, and the duration since last pregnancy or period of lactation.''

The act of suckling stimulates nerve endings that cause the anterior pituitary gland to produce prolactin. This hormone acts directly on the breasts to stimulate the secretion of milk.

Suckling also indirectly acts on the posterior pituitary gland, causing it to release another hormone, oxytocin. Oxytocin passes through the bloodstream to the breasts and acts on the small muscle cells surrounding the milk-producing ducts. This action results in the "let-down reflex" and causes the ejection of milk.

At 43, it takes me much longer to achieve an erection than it did just five years ago. Should I be experiencing such changes already?

Changes in sexual response with aging are extremely variable.

Sexual processes generally slow down with advancing age, often beginning in the forties or fifties. Like you, men in middle age can begin experiencing slower erections. They are also less likely to have spontaneous erections. Many men find that they have an increased need for direct stimulation of the penis to attain an erection.

Even though such changes are to be expected, some men, wrongfully assuming that they herald the beginning of the end, become anxious and depressed. Some may prematurely give up on sex.

But a man should be reassured by knowing that many women welcome the changes that result from his slowing down. Since sexual arousal occurs more slowly in a man of middle age or older, he is likely to engage in sexual play at a more leisurely pace—to the woman's increased gratification.

Another benefit of a man's slowing down is that he can maintain an erection for longer periods before ejaculating.

I happen to have a very small penis that measures only three inches fully erect. I have difficulty keeping it in a woman's vagina.

With some women it's even difficult for me to feel anything while thrusting. If I have to wear a condom, I might just as well roll over and go to sleep.

Do the products advertised to increase penis size work? What about surgery? I'm 26. Is there any chance my penis will grow more?

According to board-certified urologist Terrence R. Malloy, chief of the urology section at Pennsylvania Hospital, your penis is unlikely to grow. ''The anatomical size of a man's penis is generally established by the time he is 18 to 20 years of age,'' he says.

Steer clear of anything advertised to increase the size of your penis. You'd be wasting your money—and possibly endangering your health. There are no over-the-counter devices, diets, medications, exercise systems, or anything else that can safely and effectively increase the size of the adult penis. Nor is there plastic surgery for increasing penis size, as there is for increasing breast size.

However, advises Dr. Malloy, ''there are certain rare conditions which will make a penis smaller on erection than it should be.'' One of them is hypospadias. This is a congenital condition in which the urethra is mispositioned on the undersurface of the penis.

Another congenital condition that may interfere with erection is penile chordee, which is often associated with hypospadias. In this condition, a fibrous band of tissue around the urethra results in an abnormal curvature of the penis.

Peyronie's disease may also result in an abnormally curved erection. In Peyronie's, a knot of scarlike, inelastic tissue develops on the penis. The cause is unknown, and the condition usually affects men over 40.

If such conditions are present, they can often be corrected by surgery. ''Any male concerned about the size of his penis should seek a consultation before a definitive answer can be given on corrective surgery,'' says Dr. Malloy. He recommends that you seek out a board-certified urologist with considerable experience in penile surgery. A medical center near you may be able to provide a recommendation.

I'm a 53-year-old woman going through menopause. I've been getting some pretty bad hot flashes, and my doctor recommends that I get estrogen replacement therapy.

I remember reading that there were some hazards involved. Is estrogen therapy considered safe? Does it work?

Estrogen replacement therapy (ERT) can certainly help relieve your menopausal symptoms. It's particularly helpful in treating the hot flashes that so trouble you. It will also relieve the vaginal dryness (atrophic vaginitis) common to many women after menopause.

ERT is particularly effective in controlling osteoporosis, a condition in which bone mass decreases, causing bones to be more susceptible to fracture. Bone loss usually accelerates after menopause, but estrogen can retard or halt the loss. Studies have shown a substantial reduction in hip and wrist fractures in women who began estrogen replacement therapy within a few years of menopause.

You should discuss ERT with your physician. In cases such as yours, the benefits outweigh the risks.

When ERT was first used, women on long-term estrogen therapy were found to be five to eight times as likely to develop endometrial (lining of the uterus) cancer as other women. But by current standards, the dose of estrogen these women were given is considered excessive. Today, women are put on the lowest possible dose of estrogen that will relieve symptoms.

Most doctors prescribe a regimen of estrogen and progesterone. This regimen seems to have greatly reduced a woman's chances of developing endometrial cancer due to ERT.

Before you begin ERT, it's important to undergo an examination of the endometrium to detect cancerous or precancerous changes. Women should not be on estrogen therapy if they have cancer of the uterus or breast (except in some special cases where the drug is used to treat breast cancer).

Estrogen should also be avoided when a woman has undiagnosed vaginal bleeding, undiagnosed breast disease, clotting in the legs or lungs, or a history of stroke, angina, or liver or heart disease.

While you're on estrogen therapy, see your gynecologist twice a year. Report any side effects such as nausea, fluid retention, weight gain, or breast tenderness. More serious side effects may include high blood pressure, gallbladder disease, and atherosclerosis. Controversy exists as to whether or not ERT increases a woman's risk of breast cancer.

■ *Is it important for me to wear a jockstrap (athletic supporter) when I engage in sports?*

"Wearing an athletic supporter immobilizes the genitals," says Dr. William R. Bunge of San Antonio, Texas. Not only does the supporter prevent discomfort to the testicles during athletic activities, it also prevents the testicles from positioning themselves in a way that might cause injury.

Dr. Bunge especially recommends athletic supporters for men with varicocele, a swelling of the veins in the spermatic cord. The athletic supporter lightly compresses the varicocele and at the same time keeps the weight off the spermatic cord.

While an athletic supporter may make a man more comfortable for some sports, it will not protect against hard blows to the testicles. "Wearing an athletic supporter does not afford any protection from injury due to impact with a hard object, such as a baseball or a hockey stick," warns Dr. Bunge.

■ *A friend offered me some "Spanish fly" and I'd like to try it. But can it be harmful?*

It certainly can. "Spanish fly" is a dangerous chemical. Its real name is cantharis, and it's a product of a beetle found in southern Europe.

Spanish fly got its erroneous reputation as a sexual stimulant because it inflames the urinary tract and may cause an erection or an uncomfortable feeling in the genital region. It can also cause poisoning and death.

Is it true that a child raised by a homosexual father is more likely to be homosexual?

No, says Dr. Michael Carrera, author of *Sex: The Facts, The Acts, and Your Feelings*.

"More than 95 percent of the children who live with a gay parent have a heterosexual orientation," he reports. Also, remember that the overwhelming majority of homosexual men and homosexual women were raised by heterosexual parents.

CONFESSIONAL BOX

"It's not possible to have a male friend"

Linda, (37): I used to think it was possible for men and women to be just friends, but now I know that's not true. I had been friends with a certain boy since we were 10 years old. Even though we both married and had children, we kept up a correspondence. He even wrote to me from Vietnam. I thought that I could tell him anything.

Finally, my friend confessed to me that one of his major desires all these years was to see me naked or to go to bed with me. I was crushed by those words and have not been in contact with him since. I mourn the loss of what I thought was my dearest friend. I don't plan to have any more male friends.

Sometimes I worry about vaginal odor. Do you think I ought to use those vaginal deodorant sprays?

Vaginal deodorant sprays are unnecessary and potentially harmful. To combat normal vaginal odor, a woman need only wash daily with soap and water.

So-called feminine hygiene sprays are aerosols that are applied to a woman's external vaginal area. Most sprays consist of oils and perfumes, antibacterial agents, and a propellant. They work like underarm deodorants to inhibit bacterial growth and reduce the possibility of offensive odor.

Many women have allergic reactions to the sprays. Even more develop irritating skin conditions, such as rashes, blisters, itching, or irritations. The most common complaint is a burning sensation. Some women experience a feeling of urinary urgency along with pain, swelling, or open sores.

In addition to the physical problems, feminists point out that the sprays put women down by making them apprehensive about a normal part of female functioning.

Some women harm themselves because they use the spray inside the vagina when it is meant for external use only. Another problem is that the sprays may mask vaginal infections or other problems that should come to a doctor's attention.

It's best not to use such sprays. But if a woman feels they're necessary, she should follow these guidelines:

- Don't use a spray just before sexual intercourse. It can irritate a man's genitals, too.
- Don't use the spray on a tampon and then insert the tampon into the vagina.
- Don't use underarm antiperspirant on the genital area.
- Hold the spray can at least eight inches from the vaginal area.
- Use the spray sparingly, and not more than once a day.
- Don't use the spray on broken, irritated, or itching skin.

What would happen if my husband took my birth control pills instead of me? Would it be good contraception?

Birth control pills, which contain the female hormones estrogen and progestin, would have many undesirable effects on your husband.

Even in low doses, these hormones could result in bodily changes: female contours, with rounded hips; enlarged breasts; less body hair; and softer skin.

Your husband's sexual desire would decline and his frequency of erection would be reduced, making intercourse difficult or impossible. His testicles would begin to atrophy. Sperm production would decrease.

In sum, it's not a good idea.

My daughter is only nine, but she's developing breasts! Do you think there's something wrong with her? All her friends look like little girls, but she's starting to look like a woman.
What should I say to her about it?

Puberty comes unusually early for a small proportion of children.

Chances are, your daughter is going through normal development—but on a faster timetable than most girls. Since you're concerned, take the question up with her physician. In general, authorities advise, a girl should be checked by a physician if she shows signs of puberty before age eight, a boy if he shows pubertal changes before age nine.

Some young children experience isolated pubertal changes, such as the development of pubic hair or breasts. This condition is called incomplete precocious puberty. Others undergo the full range of pubertal changes—called complete precocious puberty—including maturation of the reproductive system.

Some cases of precocious puberty are so extreme that parents seek immediate medical help. Infant boys, five months old, have had penis development. Five-year-old boys have produced sperm. Menstruation has occurred in babies under a year.

The youngest documented pregnancy is that of a five-year-old Peruvian girl who began menstruating at age three. On May 15, 1939, she gave birth to a six-pound boy by cesarean delivery.

If a child shows signs of precocious puberty, chances are nothing is wrong. In most cases, doctors can find no medical cause and make the diagnosis of "idiopathic sexual precocity." This simply means that the clock of puberty is running too fast. The child will probably have to have periodic check-ups, but no medical treatment is required.

In some children, however, the precocity is caused by illness. Puberty may result from tumors in the genitals or on the adrenal gland. Congenital brain defects, a brain tumor, or an injury to the nervous system may also account for precocious puberty.

Sometimes signs of puberty occur when children accidentally ingest sex hormones, such as birth control pills or medications for menopause.

Children are likely to have a hard time adjusting to their premature development. You may find that your daughter is terribly embarrassed by her breasts and hides them in layers of loose, bulky clothing.

Experts advise parents to tell their children about any medical problems that may be causing the condition. If there is no medical cause, tell your daughter that nothing is wrong with her. Stress the fact that in a few years her friends will catch up, and then she'll feel more comfortable.

CONFESSIONAL BOX

"Now I can say 'I love you'"

<u>Ken, (28)</u>: A woman I met on-line taught me to express my feelings. I had known her on the computer for some time when we started to communicate by telephone. We became good friends, and one night she said, "I love you." Nobody, except for my Mom and female relatives, had ever said that before. Not even the girls I dated.

Looking back, I realize that I had always hoped someone would say those words to me in an honest way. I had been afraid of using them. After I said "I love you" back to her, I realized that there were several people who needed to hear those words from me, so I began telling them. I even started telling my Mom "I love you," and I haven't done that since I was a child. I still have trouble telling male friends that I love them. My Dad needs to hear it from me, too, but that will come in good time.

My wife has had herpes ever since I met her four years ago. To the best of my knowledge, I don't have the disease.

My wife's gynecologist says that if I haven't gotten herpes by this time, I must be immune to it. He says I only have to avoid sex when my wife has an outbreak. Do you agree? I love my wife dearly, but having sex makes me more than a little nervous.

Nicholas J. Fiumara, M.D., says you have probably had asymptomatic primary herpes either before or since meeting your wife. Therefore, you're immune to reinfection through marital sex.

To confirm this, you should get a neutralizing antibody test for herpes 2.

My husband developed a prolonged, painful erection a week after he started taking the antidepressant trazodone (Desyrel).

The doctor said it was most likely caused by an adverse reaction to the drug. Do you agree? Are there other drugs that can cause this strange problem?

It's certainly possible for medications to cause abnormally sustained erections. The condition is called priapism.

The Medical Letter, a reliable source of drug information, reports that trazodone may cause prolonged, painful erections even when taken in ordinary doses (100 to 300 mg/day).

Priapism is also associated with anticoagulants, corticosteroids, tolbutamide (Orinase), methaqualone (Quaalude), chlorpromazine (Thorazine) and other phenothiazines, and the antihypertensives guanethidine (Ismelin), hydralazine (Apresoline), and prazosin (Minipress).

Priapism is a medical emergency. It can lead to permanent impotence if it is not treated quickly. If a man develops an inappropriate, painful, or prolonged erection, he should discontinue the suspected drug and notify his physician immediately.

When I have intercourse, I sometimes have stomach cramps. What causes this? What can I do to stop it? I'm a 20-year-old male.

Randall Rissman, M.D., a family physician, advises that you take note of your position next time the cramps occur. You may be assuming an awkward position that strains the abdominal muscles and causes them to cramp.

Another possibility is that you've been having intercourse late at night after a heavy meal. "Many people experience abdominal distress if they lie down after a meal," says Dr. Rissman. "If they've had alcohol or caffeine, they may be even more uncomfortable."

If you're anxious about sex, you may experience cramps, too. Dr. Rissman notes that "the gastrointestinal tract is a favorite target for anxiety."

If the cramps also occur at other times and persist, see your doctor. Abdominal cramps are a symptom of many conditions.

I've heard that if you have sex before 18, you have a higher chance of getting cancer of the cervix. Is this true?

Yes. Several studies have shown that women who begin having sexual intercourse at an early age with multiple sexual partners are at a higher risk of developing cervical cancer than their peers who delay intercourse and limit the number of partners. There is a virtual absence of cervical cancer among nuns, while a high prevalence is found among prostitutes.

In addition, genital herpes, which has reached epidemic proportions in this country, is considered to predispose women to cervical cancer. All of these facts add up to a good reason for young women to abstain from early sexual intercourse with many partners.

To detect cancer of the cervix, women should have a Pap smear about once a year after beginning sexual activity. This virtually painless, inexpensive test can detect cervical cancer in its early stages, when the disease is almost always curable.

I'm a 26-year-old "man." But all my life I've felt that I was born into the wrong body. I know I must bring my body into alignment with my female soul.

I'm interested in surgery because I want to have a female body to make me a whole person. I'm not a homosexual, and I don't wear women's clothing. I am a woman in the wrong body, and I find no comfort in this world.

Where can I go for help?

You may, indeed, be a person who could be helped by transsexual surgery, the so-called "sex-change operation." But bear in mind that changing one's sex is a lengthy, costly, and difficult process. It will require a strong commitment on your part.

Your first step is to seek out a psychiatrist or psychologist who has had experience in working with transsexuals. He or she will want to do a complete psychiatric evaluation before hormone therapy and/or surgery are recommended.

That would be in line with the standards of care recommended by the Harry Benjamin International Gender Dysphoria Association. These guidelines, which are usually followed in such cases, also require that you be recommended for surgery by at least one psychiatrist and another professional who may be either a psychiatrist or a psychologist. One of these experts has to have known you for six months or more.

A careful evaluation eliminates people who are seeking surgery because of depression, schizophrenia, or other problems other than transsexuality.

If you're found to be a candidate for transsexual surgery, you will be referred to an endocrinologist or a gender identity clinic. Hormones are generally prescribed for a six-month trial period during which you decide whether to go ahead with the surgery. If you decide not to, discontinuing the hormones will slowly reverse the physical changes that have taken place.

When males are given female hormones prior to surgery, they can expect the following changes: a softer skin texture, an increase in hair growth on the scalp, diminished muscular development, a more feminine distribution of fatty tissue, and increased breast size.

Hormones, by the way, will not make the voice higher. Nor will they do much to reduce beard growth. So a male undergoing hormone therapy needs to have his beard removed by electrolysis over a one- to two-year period. At least half of this process must be completed before surgery is recommended.

Most clinic programs and individual physicians insist that the transsexual have some supportive therapy. An understanding counselor can be quite helpful during the transition period. Counseling is recommended for some months after surgery, too.

Before surgery, most programs require the candidate to live the life of the new sex. That means a male who wants to be female must dress and work as a woman for six months to two years. Having actually experienced his new role, he's better able to make a final decision about the surgery.

In male-to-female sex reassignment surgery—which is sometimes performed in two stages—the male organs are removed. Then, a vagina, clitoris, and labia are constructed. Usually, surgeons use the skin of the penis and scrotum to construct the female anatomy. Sometimes skin grafts from the buttocks may also be used. Silicone implants may be surgically inserted to increase breast size.

Although they have female genitals, male-to-female transsexuals are not able to bear children. They have none of the internal female sex organs required for pregnancy.

Erotic sensation after surgery varies. Sexual feeling depends mainly on the degree to which sexual nerves have remained intact. Almost all male-to-female transsexuals report the presence of some erotic feeling. For some, the feeling of orgasm is as intense as before. Others experience a more diffuse sexual sensation.

For more information about transsexualism (and transvestism), write to The Janus Information Facility (1952 Union Street, San Francisco, CA 94123), which publishes such pamphlets as *Information for the Family of the Transsexual* and *Legal Aspects of Transsexualism.*

I've heard that there's some medicine that can be given to a woman after a rape to prevent pregnancy. What's it called? Does it really work?

There is indeed medication that can prevent pregnancy after a rape. It's intended only for such emergencies and is not safe as routine ''morning-after'' contraception.

Physicians advise that within seventy-two hours of the attack the woman be given two tablets of 0.5 mg norgestrel and 0.05 mg ethinyl estradiol (Ovral). She should get two more tablets twelve hours later. Menstruation should occur within two to three weeks.

Another preventive is insertion of a Copper-T or Cu-7 IUD within five days of the attack.

Can semen be frozen or refrigerated? If it is, does it lose its potency? How long will it keep?

Yes, semen can be frozen.

Before having a vasectomy, some men choose to have their sperm frozen in sperm banks. They hope that if they change their minds and want to have more children, the defrosted sperm will be able to produce a pregnancy. Thawed sperm have, indeed, resulted in hundreds of healthy babies. Some of them were born more than ten years after the sperm was frozen.

In general, however, frozen sperm declines in its ability to cause pregnancy. The longer the sperm is frozen, the less likely it is that pregnancy will result.

I've been married for over two years. My wife and I have very little sex because she says it always hurts.

I try to take my time and make sure that she is ready. But I think she must be afraid because nothing helps. I'm almost ready for a divorce. What can we do?

Your wife should consult a gynecologist. There are many physical reasons why a woman might find intercourse painful.

Vaginal infection is a common cause of painful intercourse. The infection may precipitate a burning sensation that persists after intercourse. Some venereal diseases produce sores that may make intercourse painful.

Women can also find intercourse painful because of allergic reactions to chemical contraceptives, such as foams or suppositories. Some women are allergic to douche preparations or to the rubber in diaphragms or condoms. An allergy to feminine hygiene sprays or bubble bath may irritate the genitals and make intercourse uncomfortable.

Intercourse can become painful when a woman's vagina does not adequately lubricate because she is not sufficiently aroused. But inadequate lubrication may also be due to medication she is taking, such as antihistamines. Also, after menopause, vaginal lubrication often decreases. The use of a lubricant such as K-Y Jelly or prescribed estrogen cream usually solves the problem of insufficient lubrication.

Intercourse may also be painful because of congenital abnormalities or tumors of the genitals. Hymen problems, such as a rigid or thick hymenal ring, sometimes account for uncomfortable intercourse.

A woman who suffers from rectal or intestinal problems may find sex painful. So may a woman with a urinary tract infection. Pelvic inflammatory disease or cysts on the ovaries may cause a woman to feel lower abdominal pain induced by the man's thrusting.

Another cause of pain during intercourse may be endometriosis. In this condition, tissue from the uterine lining occurs in abnormal places. Deep thrusting usually causes pain. The pain is usually most severe just before menstruation.

After childbirth, a poorly healed episiotomy may cause a sharp pain at the vaginal opening.

Any irritation, scrape, or cut on the vulva or vagina is likely to result in painful intercourse. Abscesses often cause sharp, localized pain, with burning or searing.

When women suffer from vaginal spasm, penetration may be difficult or impossible. There is often pain at the vaginal opening as the man attempts penetration. The condition is usually thought to be psychological in origin, but it may also be related to previous vaginal trauma, such as surgical scarring.

As you suspect, psychological factors can contribute to painful intercourse. But your wife should check out all of the physiological reasons before she attempts to get psychological help.

I recently spent the night at a friend's house. (We're both 14-year-old boys.) We both heard each other masturbating, so we decided to masturbate together. I've been worried about this ever since.

I still like girls, but I really liked it when we rubbed each other's penises. Do you think this means I'm gay?

You're probably going through a stage of psychosexual development that's widespread but little discussed.

It's extremely common for boys and girls to engage in sexual experimentation with members of their own sex. Adolescents are often in a state of high sexual tension that can be discharged with either sex. Sometimes a member of the other sex is not available. And, even if one were accessible, sexual involvement might be premature in terms of the person's overall development.

Obstetrician-gynecologist James P. Semmens, of the University of California College of Medicine, considers same-sex sexual experiences normal responses to adolescence and puberty. They are, he says, part of maturation toward heterosexual relationships in later life.

Among adolescents, the type of incident you describe is rarely an indication of homosexuality, which is a basic sexual and emotional affinity for one's own sex. Says psychiatrist Ellen Rothchild of Case Western Reserve University: "What passes for homosexual behavior in childhood and adolescence is often a normal part of growing up, and does not foretell a future homosexual orientation."

Teenagers, however, almost never talk with one another about same-sex erotic experiences. So it's no wonder that many young people believe they are gay when they have same-sex contacts.

In addition, many adolescents never actually engage in homosexual play, but would like to—and are disturbed by their desires. They may have fantasies about engaging in sexual activity with their best friends, for example.

Their distress arises from a mistaken belief that fantasy necessarily indicates heterosexuality or homosexuality. In fact, fantasy often has a life of its own. Usually, it's normal to have such fantasies. Heterosexuals do not only feel attracted to members of the opposite sex.

My wife is convinced that she got pregnant by moving her bowels after we had sex. She believes this dislodged her diaphragm.
Is this possible?

Indeed, it is. A normal soft stool isn't likely to have this effect, but a large, hard stool can displace the rim of the diaphragm.

This displacement is especially likely to happen if the diaphragm is loose. It may also occur if the woman has a rectocele, an abnormal bulge from her rectum into her vagina.

A woman would be well advised to defecate before inserting a diaphragm. It's also a good precaution for her to check the fit of the diaphragm with a gynecologist at least once a year. She should be especially sure to check the fit if she's gained or lost twenty pounds or more.

My doctor says I have a vaginal infection called trichomoniasis. How did I get it? Is it serious?

Trichomoniasis is a vaginal infection caused by a protozoan. It's usually spread through sexual intercourse. The organism may occur in the gastrointestinal system, however, and can be carried from the rectum to the vagina by anal intercourse followed by vaginal intercourse or by wiping bowel movements from the rectum toward the vagina.

Since the protozoan that causes trichomoniasis can survive outside the body in a warm, moist place, you may also have contracted the infection through clothing, toilet seats, towels, washcloths, and other personal articles. It may also be transmissible through communal hot tubs.

The infection is usually easily treated with metronidazole (Flagyl). A male partner needs to be treated simultaneously. Men often harbor organisms in the urethra, prostate gland, or bladder without any symptoms, and thus can repeatedly reinfect a female partner who has been treated.

As for the seriousness of trichomoniasis, its symptoms—usually a foul-smelling discharge and burning around the vagina—are uncomfortable but not alarming. However, recent evidence shows that women with a history of trichomoniasis run the risk of becoming infertile due to scarring and inflammation of the fallopian tubes.

The infection can invade the tubes without showing any symptoms, yet it may cause enough damage to make conception difficult or impossible. To help guard against this possibility, take all the medication your doctor has prescribed. Go for follow-up visits, if your doctor suggests them.

Also, be sure to wipe bowel movements from your vagina backwards. Make certain your partner uses a condom if you engage in anal intercourse. And see your doctor at the first sign that the infection may be recurring.

I regard myself as an emancipated woman. But if I indicate a desire for sex, men seem to get scared and shy away. Why?

A man may be scared because he fears that you won't tolerate his lack of knowledge or experience. He may have fears about intercourse in general. Or he may be anxious about having intercourse with a person he envisions as threatening and aggressive.

George L. Ginsberg, M.D., a psychiatrist at New York University School of Medicine, suggests yet another possibility. He says that a man may not be scared but may nonetheless shy away from accepting a sexual invitation because:

- He's not sexually interested in a particular woman.
- He's not interested at that time.
- He wants to be the one who initiates sexual contact.

You might want to consider whether you're coming on too aggressively. Cautions Dr. Ginsberg: "What is reported as a simple indication of a desire for sex may, in fact, be an aggressive gesture on the part of the woman of which she is unaware."

CONFESSIONAL BOX

"I'm inept at finding the right person"

<u>Ann, (39)</u>: I can sympathize with the guy who fears he won't have any children. I, too, am 39 and very worried that I will be childless. I wonder if it isn't even worse for a woman. With each breakup, it becomes harder and harder for me to come out of the depression.

There must be some man out there who has the same goals I have, but the problem is finding him. I keep finding men who say that I'm nice but not the one they want to settle down with. The last three guys I've dated have married another woman after breaking up with me. Since I was a "wallflower" in my younger years, I feel inept at the art of finding the right person. And my feeling of terminal ineptness increases with age!

To the guy who wants children, I'd say go for it. Don't regret twenty years from now that you didn't get what was in your power to get. Peace.

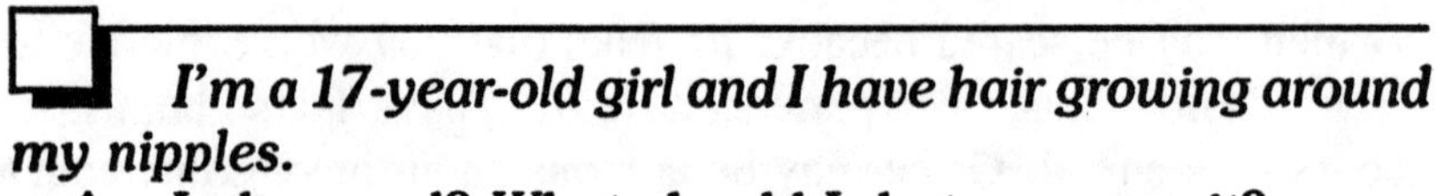

I'm a 17-year-old girl and I have hair growing around my nipples.
Am I abnormal? What should I do to remove it?

Hair growth around the female areola—the dark ring around the nipple—is normal and common. It's usually a cosmetic rather than a medical problem.

If the hair is objectionable, electrolysis can remove it permanently. Some women bleach dark hairs with hydrogen peroxide. If there are just a few dark hairs, they can be plucked with tweezers.

To be on the safe side, see a physician. If the hair has grown abruptly—and if there is a considerable amount of it—your doctor will want to rule out hormonal disorders that could be causing abnormal growth.

My 14-year-old son doesn't want to go to the pediatrician for his physicals any more. Should I take him to our family doctor? And should I go into the examining room with him? (He's become very modest in the past year.)

Like many adolescents, your son considers himself too old to sit in a pediatrician's waiting room with infants. What's more, many teenagers have sex-related concerns that they're too embarrassed to discuss with a doctor who's known them since infancy.

At the same time, few teenagers have formed an independent relationship with a private physician. Your own family physician may be a good choice. On the other hand, if your son would prefer seeing a doctor who is not closely associated with you, a specialist in adolescent medicine would be a good alternative. This is a new specialty that recognizes adolescents as forming an age group with unique medical and psychological problems.

Says one such specialist: "The only real difference between us and other physicians is that we are trained to see the teenager or young adult as a special person—a patient with problems that reflect the turbulent physical and emotional changes that take place during the adolescent years."

Ask your child's pediatrician or your own physician for a referral, or call your local medical society. It's easier to find an adolescent specialist if you live in a big city. Another good choice for your son might be an adolescent clinic. These are usually connected to large hospitals.

A sensitive doctor, whether adolescent specialist or family physician, will spend some time trying to make your son feel at ease by chatting with him before or after the examination. Sometimes it may take several sessions to establish a good relationship between doctor and adolescent.

Your son needs to know that everything he says to his doctor will be kept secret. Reassure him that the physician-patient relationship is confidential and that the doctor will repeat nothing to you.

Reinforce your son's independent relationship with the physician by respecting his privacy. You should not be in the examination room with your child. Nor should you inquire closely about what went on in there.

After the physical, the doctor will most likely discuss the results with you, while respecting your youngster's expectation of confidentiality about specific areas.

Most doctors will speak to you in your child's presence. By doing so, the doctor demonstrates that he leveled with the adolescent when he assured him of confidentiality. Talking to the two of you together also forestalls any misunderstanding about the diagnosis and the doctor's instructions.

I'm a 35-year-old man who sometimes goes out dressed as a woman. Is there a legal danger in doing this?

The Gateway Gender Alliance, which provides information and support services for transvestites and transsexuals, suggests that you consult an attorney about the laws in your area.

In many places, it is indeed a violation of the law to publicly wear clothing that is normally reserved for the opposite sex (although women wearing male attire are quite unlikely to be stopped). In other areas, it is not unlawful to cross-dress either in public or in private. States that do not have laws against cross-dressing include

Alabama, Alaska, Arizona, Arkansas, California, Colorado, Florida, Georgia, Hawaii, Idaho, Illinois, Kansas, Maryland, Michigan, Montana, Nevada, North Carolina, Ohio, Oregon, Oklahoma, Pennsylvania, South Carolina, Texas, Virginia, Vermont, West Virginia, Washington, and Wisconsin.

Even if a state does not prohibit cross-dressing, however, a local ordinance may forbid it. And even if there are no local ordinances, law enforcement officials may "be personally offended by the act of cross-dressing and may pick you up on one of the many 'Catch-22' laws available to them," says the Gateway Gender Alliance.

The Alliance advises you to be courteous and polite if you're arrested while cross-dressing. You're not required to provide information other than your name, address, birth date, and social security number. You have the right to call your attorney.

Don't volunteer to tell the police where you work. They may tell your employer of the incident, which may, of course, lead to termination.

The Alliance urges transvestites to do nothing to invite attention from the police. "Don't be a public nuisance, don't be drunk in public, don't walk alone in dimly lighted areas late at night. If you must go out alone at night, have a dog you can take with you—walking a dog before bedtime is a good excuse and a 'normal thing to do.' "

I seem to have an odor from my genital area. My physician assures me that I don't have a vaginal infection or a venereal disease.
What could account for the unpleasant smell?

The odor could be caused by smegma of the clitoris, a cheesy, odoriferous secretion of the sebaceous glands.

Most physicians are aware of the need to inspect the male foreskin for accumulation of smegma and the inflammation it can cause. But little attention is given to the importance of cleaning a similar structure in the female: the clitoris and its hood.

Smegma can cause inflammation of the clitoris. It may also cause adhesions that make sexual manipulation of the clitoris painful.

Other causes of odor from the vulva include perspiration, urine, or fecal matter. Odor may be compounded by poor ventilation caused by tight underclothing.

To avoid odor, wash the entire genital area with soap and water as part of your morning and evening clean-up. Be sure to wash the perineum (the area between the vagina and the anus). Take care to wash under the clitoral hood.

Even a woman who practices excellent hygiene may tend to overlook the vulva. But remember that no matter how clean you are, you can't expect to have a complete absence of scent.

Does masturbating about once or twice a day affect your penis growth? I have been losing penis growth lately. I'm 15.

There is no evidence that masturbation has any effect on penis size.

What might give you the idea that your penis is getting smaller? You may have noticed that your penis gets slightly (and temporarily) smaller in a cold bath or shower, or in cold weather. Some men also find that fatigue or emotional stress can cause the penis to shrivel up a bit.

It's also possible that you've experienced a recent adolescent growth spurt, making your penis seem somewhat smaller by comparison. Gaining a lot of weight might have produced the same effect.

I am fighting my desire to have sex with children. So far, I have only exposed myself to them while riding around in my car. But the urges are getting stronger all the time.
What can I do?

Seek professional counseling immediately. Go to a psychiatrist, mental health center, hospital clinic, or social service agency. Or call a crisis center hotline (check the yellow pages of your telephone book) and ask for referrals. Professionals are available to help people with your problem.

Don't delay in getting treatment. Sexual abuse of a child is not only illegal but may also have long-lasting emotional consequences for the victim. Remember, too, that child molesters get very little sympathy from jurors.

It's impossible to say why you have an urge for sexual contact with children. This is a matter you must discuss with a therapist. Researchers have found, however, certain common characteristics among people who seek to have sex with pre-adolescents.

For the most part, they are men. Often, they merely want to look at the child nude. Sometimes the child's genitals are inspected, touched, or kissed. In other instances, the child may be required to masturbate the adult or perform fellatio. It's rare for molesters to have sexual intercourse with their victims.

Most molesters entice their victims with favors, gifts, or promises. Some youngsters are intimidated or threatened. But outright brutality is not usually involved.

Most child molesters share a sense of alienation. They feel inadequate and inept in personal relationships. They tend to see themselves as helpless victims.

The typical child molester is a married man with marital problems. He often has trouble functioning sexually with his wife. That's why he may achieve a sense of sexual adequacy through sexual contact with a child.

Typically, a child molester commits the same type of offense over and over again. The children he seeks out are all of the same age, sex, and general physical appearance. Sometimes the molester uses alcohol to lower his inhibitions so that he can commit the offense.

Many types of therapy have been used to treat child molesters. Success has been reported with hypnosis, psychotherapy, and behavior modification.

What does semen consist of? Is it harmful to swallow? How many sperm are there in the average ejaculate?

Semen consists of fluids supplied by the prostate, seminal vesicles, and Cowper's glands. The prostate provides the greatest portion of the seminal fluid.

Semen is also composed of ascorbic and citric acids, water, enzymes, fructose sugar, phosphate and bicarbonate buffers, and a variety of other substances. None of these materials is harmful if swallowed. But infectious organisms can be transmitted through swallowed semen.

The semen of a single ejaculate—about one teaspoonful—typically contains between 200 million and 500 million sperm!

Semen also contains a natural antibiotic substance, an enzyme called seminal plasmin. This protein is, apparently, at least as potent as penicillin. It's been found to destroy staphylococci, streptococci, bacilli, and other microbes that can cause illness.

The German researchers who discovered seminal plasmin believe that intercourse may be a natural way to prevent some types of vaginal infection.

The amount of semen a man ejaculates depends on a variety of factors including his age (older men tend to produce less fluid) and the length of time since his last ejaculation.

The consistency of semen also varies from man to man. Some men ejaculate a fluid that is thick and almost like gelatin. Another man's ejaculate will be thin and somewhat watery. In addition, one man's semen may vary in consistency from ejaculation to ejaculation.

 I'm soon due to leave the hospital after recovering from a heart attack. My doctor has told me nothing about resuming sexual activity.
What's recommended?

You should initiate the discussion with your doctor—with your sex partner present. Ask your doctor what kind and amount of sexual activity is safe for you at this stage of your recuperation. Unfortunately, you're not alone in trying to get such information. Sex researchers have found that two-thirds of patients who suffered heart attacks were given no sexual advice at all during their medical treatment.

Coronary patients often fear that sexual activity will bring on another heart attack. Many people unnecessarily abstain from sex for this reason. The result can be frustration and marital conflicts that may make the heart condition worse.

In general, physicians find that less-taxing forms of sexual activity are a good way to begin sexual rehabilitation. Such activities can boost your confidence and ease the resumption of intercourse.

When you first come home from the hospital, you and your partner can engage in pleasuring—touching, cuddling, and stroking. This kind of gentle activity should be done without the expectation of intercourse. It helps relieve anxieties that can result in impotence.

Most heart attack patients can safely go on to masturbation or oral sex as soon as they can tolerate an accelerated heart rate of 130 beats per minute. In an uncomplicated recovery, this usually takes four to eight weeks.

Cardiologists find that intercourse requires much less physical effort than is popularly believed. Heart attacks during intercourse are extremely rare. In fact, fighting a traffic jam or having an argument can be more dangerous to a person who's had a heart attack.

You can usually resume intercourse when you're able to climb one or two flights of stairs or walk several blocks at a brisk pace. For the average person, that means about four months after the heart attack.

Three positions that lessen cardiac workload are recommended for the postcoronary patient. These are:

- Lying on the side, in a face-to-face position
- Lying on the back, with the partner on top
- For men, sitting on a wide chair, low enough for the feet to touch the ground.

It's also prudent to have intercourse in a cool, dry room. Heat and humidity can increase the strain on your heart. Also refrain from intercourse after eating a big meal. Digesting the meal can tax your circulatory system.

It's safest for married people to have intercourse only with their spouses. Sex outside of marriage, especially if it follows heavy eating and drinking, can be extremely stressful.

Stop sexual activity if you feel chest pain. Tell your doctor about the pain and other danger signals, such as palpitations that continue for fifteen minutes or more after intercourse and marked fatigue on the day after intercourse.

I'm a 19-year-old virgin male who masturbates frequently and often fantasizes during masturbation.

Lately, I've been having sexual dreams more often than usual. These dreams disturb me because, in them, something always stops me from having sex. I'm always a virgin in my dreams.

Are these sorts of frustrating dreams normal? Will I ever be able to have intercourse in a dream before I have a real-life sexual encounter?

Harold I. Lief, a psychiatrist, says that your dreams are not unusual. Often, young men who have not yet had intercourse will stop short of intercourse in their dreams. These dreams do not necessarily mean that you are inhibited about intercourse itself.

You may be able to help your dream life along by fantasizing about what intercourse might be like. If you can develop a mental image, it may help toward having the experience in your dreams.

To determine whether you are inhibited about sex or whether the frustration you are experiencing is normal, watch for feelings of anxiety as your sexual scenario unfolds. If you can fantasize about sexual intercourse during masturbation without anxiety, you'll eventually be able to complete the process in your dream life.

My wife and I are both in our early forties and we have three children.

Our family is complete, and we're considering sterilization. Which is safer and more effective, male or female sterilization?

Vasectomy—the male sterilization—is as effective as female sterilization, but safer.

According to a study of the medical literature published in the *American Journal of Public Health*, there have been no reports of death from vasectomy in the United States. In contrast, for every 100,000 female sterilization procedures—including both laparoscopy and laparotomy—there are 4.72 deaths from laparoscopy and 2.29 deaths from laparotomy.

The study also found that vasectomy has fewer complications than female sterilization procedures. Of 100,000 vasectomies, only 43 resulted in major complications. But out of 100,000 female sterilizations, 6,170 laparotomies and 2,100 laparoscopies were marked by major complications. These included hemorrhages that needed transfusions and problems that required the use of intravenous antibiotics.

Vasectomy is also likely to be less expensive and to require less recuperative time than female sterilization.

I'm a 27-year-old woman. I seem to be getting yeast infections very frequently.

The medicine I take is effective for a few weeks. Then the infection comes back. Is there anything I can take to get rid of it once and for all?

How can I avoid getting yeast infections in the future?

Recurrent yeast infections require a prolonged, varied attack. This is because resistant strains of the infecting organism—*Candida albicans*—are normally found in the vagina.

Armando DeMoya, M.D., an obstetrician-gynecologist, reports

good results with a ten-day regimen that includes the application of miconazole nitrate (Monistat cream) into the vagina, plus the insertion of tampons impregnated with gentian violet (Genapax).

The tampons are inserted before bedtime and on getting up in the morning. In addition, the vulva and vaginal areas are painted with gentian violet once a week for three to four weeks.

For resistant infections, Dr. DeMoya also recommends thrice-weekly douching with yogurt (one tablespoon of active culture in one quart of warm water) for a period of three weeks.

To promote circulation of air, patients should temporarily refrain from wearing underpants. Instead, they can wear crotchless pantyhose or stockings with a garter belt. Similarly, jeans or slacks should be avoided during treatment.

Susceptible women should always avoid nylon undergarments and tight pants, particularly in warm weather. They should be aware that they'll be prone to the condition during periods of illness or emotional stress, when the body's resistance to infection is lowered. Taking oral contraceptives or antibiotics can also increase the chances of getting a yeast infection.

Since the infection often occurs in association with diabetes and Addison's disease, those conditions should be checked for in women with recurrent yeast infection.

I'm a 23-year-old guy. My roommate's girlfriend has herpes, and she sometimes takes showers at our place. Can I get herpes from using the same towels?

Evidence seems to indicate that genital herpes is transmitted only through sexual contact. However, laboratory research shows that the herpesvirus can survive for hours on surfaces outside the body, such as towels, toilet seats, plastic benches near hot tubs, and clothing.

That means there's a theoretical possibility that herpes can be spread in ways other than through direct contact. Although your risk of contracting herpes by sharing towels is not very great, it would be prudent to avoid sharing such personal articles with your roommate's girlfriend.

I've heard about some sort of pill that can be inserted into the vagina to cause an abortion. Why isn't this used more, since it sounds so easy?

You're probably thinking of prostaglandin suppositories.

Prostaglandin is a hormonelike substance that causes uterine contractions. When it's inserted into the vagina, a prostaglandin suppository can indeed cause labor and miscarriage.

This newest, least-known method of abortion, which must be performed in a hospital, is not yet in common use. Many hospitals use it only when the fetus has already died but the woman has not spontaneously gone into labor to expel it.

Common problems with prostaglandin suppositories include nausea, vomiting, diarrhea, fever—and failure to cause abortion. Until these problems are remedied, you're unlikely to see the drug in general use as an abortion method.

I'm a 15-year-old girl, and I have mono (mononucleosis). My friends are teasing me, because they say you get it from kissing. But I didn't kiss anybody. How did I get it?

Mononucleosis (also called infectious mononucleosis or glandular fever) is sometimes called the "kissing disease." But how it actually spreads remains a mystery. The incidence among sexually active people is no higher than among those who are not sexually active.

Although mono is an infectious disease, it's only slightly contagious. It's rare that everyone in a household comes down with it, and it usually doesn't even spread between husbands and wives.

Mono is caused by a virus and is thought to be spread through airborne droplets or through direct oral contact. But it has not proved possible to spread the disease experimentally, even by spraying mono patients' saliva into the throats of volunteers.

So you can tell your friends that nobody knows exactly how mono is spread, but that your case was definitely not contracted through kissing.

Mono is primarily a disease of the teenager and the young adult.

Some 75 percent of cases occur between the ages of 15 and 30.

Symptoms vary, and may include a sore throat, stiff neck, abdominal pain, fever, rash, swollen glands, and a general feeling of fatigue. The disease is usually mild, but it may linger. Acute symptoms may last a week or two.

Mono can be detected by a three-minute blood test. The usual treatment is bed rest and convalescent care. Recovery takes from two to six weeks.

■ *I find it very sexually stimulating when my wife plays with my nipples.*

Is this unusual? Do other men have sensitive breasts?

Many other men have sensitive breasts, too, although most heterosexual males have little experience with breast eroticism—it's more likely to be practiced among gay males.

Noted sex researcher Alfred Kinsey found that ''a few males may even reach orgasm as a result of breast stimulation.'' In fact, as many males—both heterosexual and homosexual—may be sensitive to breast stimulation as females.

■ *What's the best way for me to act when I see a man exposing himself on the street or in the subway? I'm concerned because this sometimes happens when my five-year-old daughter is with me.*

Exhibitionism (indecent exposure) is best handled by pointedly ignoring it. If you see a man exposing his genitals in public, pretend that nothing out of the ordinary is happening.

Exhibitionists (or ''flashers'') are almost always men. Their targets are almost always women or girls they do not know. The majority of exhibitionists are not dangerous or violent. Nor are they usually potential rapists.

About half of all exposures take place in the front seat of a car. The exhibitionist calls a woman over on the pretext of asking directions. As she begins to answer, she notices that his penis is exposed. Generally, he's staring at her, waiting for a reaction.

Although most exhibitionists are not dangerous, some are. Psychiatrist Richard T. Rada of the University of New Mexico School of Medicine found that some rapists had a history of exhibitionism before they engaged in rape. Sometimes the rape occurred when the man was intending only to expose himself. Advises Dr. Rada: "The safest approach for the victim of an exhibitionist is to view all sexual deviates as potentially dangerous."

Thus, it's wise not to give the exhibitionist the response he's looking for. Psychiatrist-lawyer Robert L. Sadoff of the University of Pennsylvania School of Medicine notes that "ignoring the exhibitionist will be disappointing to him but will not provoke him to action or violence."

You should particularly avoid exhibitionists who shout obscenities or make threatening gestures. Also be careful to say or do nothing that may humiliate or anger the exhibitionist. If he attempts to block your way—especially when you're alone at night—scream loudly and walk around him.

"Screaming is a shattering experience for the exhibitionist," observes psychiatrist Thomas P. Hackett of Harvard Medical School. "I know some who gave up exposing themselves for many months following an encounter with a screaming woman."

Tell the police about a man who repeatedly exposes himself, on the same street corner every day, for example. The police can direct him toward psychological counseling, perhaps even without bringing formal charges against him.

It's rare for exhibitionists to expose themselves to children. If your daughter happens to be with you when a man exposes himself, tell her to ignore him and just walk by. Assure her that there's no reason to be frightened. You can explain that most exhibitionists are timid, frightened souls who cannot make contact in any other way. Dr. Hackett says that, in his experience, children are not as frightened by exhibitionism as adults are.

You must remember, though, that a small percentage of exhibitionists will approach children sexually. If you witness an act of exhibitionism directed specifically toward a child, notify the police.

Most exhibitionists are between the ages of 15 and 35; they have an average level of intelligence and education. They're usually employed or in school and have no police record. Many are married and some are parents.

Most exhibitionists are inhibited. They have not known many women and, for them, sex is a guilt-laden subject. If they're married, they are often troubled by impotence, premature ejaculation, and sexual insecurity. The typical exhibitionist is a passive person who feels insignificant and inadequate.

The primary motive of the exhibitionist is usually not sexual gratification, even though he may have an erection and be masturbating. Actually, his act gratifies emotional needs. For example, an exhibitionist may expose himself when something threatens his precarious sense of self-esteem, making him feel anxious.

The anxiety is partially dissipated by the exposure, notes psychiatrist James L. Mathis of the Medical College of Virginia. Any sort of reaction from the female reassures him that he does have a penis and is really a man.

Dr. Hackett sees exhibitionism mainly as an expression of anger. The exhibitionist exposes his penis instead of expressing direct hostility toward a dominant mother and a weak or absent father— a common combination in the background of exhibitionists.

The exhibitionist rarely seeks medical attention until he's forced to. Even then, he may deny the problem. That's why it's not easy to treat an exhibitionist. Neither drugs nor jail sentences have proven to be effective.

"About the only thing I have known to work is psychotherapy," comments Dr. Hackett. First, an exhibitionist must admit to the problem. But many of those who do, get good results from psychotherapy within six months. Both group and individual therapy may be effective.

I'm a 21-year-old woman, recently married. I find myself wondering about making love in places other than my bedroom. Is this normal?

It's normal, not only to fantasize, but also to make love in a variety of places. Many people feel guilty or wonder if they're "kinky" if they want to experience something that they consider out of the ordinary.

A general guide is that anything which takes place between consenting adults, out of the sight and hearing of others, is permissible.

This would rule out having sex in a public place where it might offend other people. But there is a great variety of appropriate private places for sex.

My gay lover has recently been diagnosed as having AIDS. What form of sexual activity can we still have? I'm assuming that condoms are safe.

You must avoid the exchange of semen through either oral or anal intercourse. Don't assume that condoms are totally safe, although they may provide some protection. The Centers for Disease Control say that ''the consistent use of them may reduce transmission,'' but theoretically, it's possible for viral particles to get through the pores in a condom.

The safest forms of sexual activity for you as a couple are massage, snuggling, and mutual masturbation.

Fred Westendarp, M.D., also advises you to avoid the exchange of body fluids by not sharing shaving equipment or toothbrushes. Since there is debate about the role of saliva in the transmission of the AIDS virus, you might want to give up intimate kissing until more is known.

It's also important to keep your general health in top condition so that you do not bring an infection home to your partner. Get plenty of rest. Eat a well-balanced diet and exercise regularly. Make sure your immunizations for diseases such as hepatitis B and flu are up to date.

For more information about coping with AIDS, contact the Gay Men's Health Crisis Hotline, 254 West 18 Street, New York, NY 10011 (212-807-6655).

I'm taking medication for depression that has the effect of lowering my sex drive.

This is causing marital difficulties. I don't want to stop taking the medication. But I'm worried that my wife will find someone else. I'm 39 years old.

Antidepressants—such as Tofranil, Elavil, Vivactil, and Norpramin—can indeed interfere with sexual performance. Men taking such medication may experience delayed ejaculation, failure to ejaculate, and difficulties in attaining or maintaining an erection.

It's important for you to let your doctor know that you are having sexual difficulties. Your physician may be able to minimize or eliminate the problem by prescribing another medication or by altering the dosage of the one you're taking.

Your doctor should also explore whether your problems may be compounded by other factors, such as alcohol, social drug use, marital problems, or illness.

I'm a 32-year-old man. Although I'm not gay, I constantly fantasize about spanking a teenage boy. When I see a young man on the street, the first thing I look at is his bottom, and I start to imagine him in various positions, waiting for a strap or paddle.

As a youngster, I was aroused by tales of my friends' "whippings." I, myself, was rarely spanked. However, the threat was there in the form of a razor strap hanging on the pantry door. I felt it on my own backside only a dozen times or so, and I didn't like it.

I'm not married, but I enjoy having sex with females. Sometimes, while making love, my mind wanders off to young men being spanked and I find myself virtually masturbating into the woman's vagina. It makes me feel as if I'm using her.

Do any other men have this preoccupation? What can I do about it?

Sidney Rosen, M.D., a psychiatrist, reports that your fantasies are relatively common. There may be all kinds of ''reasons'' for them. Perhaps you're identifying with your father, who had the power to whip you. Unconsciously, you may associate that power with being an adult male.

At the same time, your interest in your friends' whippings was connected with the development of your own sexuality. A certain number of people tend toward masochistic preferences—they are aroused by pain. Apparently, your masochism was once removed: You preferred not to experience the pain, but to have it happen to someone else.

There is nothing wrong with using your fantasies of spanking (or anything else) to heighten your excitement during sex. If you would like to have closer contact with your partner, though, you can train yourself, gradually, to transfer your attention from these fantasies to her body or her responses. Such a transference will result in more mutuality in your sexual experiences.

You may be able to do the transferring by yourself, but professional help would make it easier.

CONFESSIONAL BOX

"I was a child porno star"

<u>Ted, (42)</u>: I was molested from age six to 15, at which time I ran away from home. The type of life I led before I ran away was unusual to say the least. My father was in an underground pedophiliac organization that used to swap and photograph young boys. I was a child porno star. Today, I'm going through psychotherapy for what I experienced as a child. In therapy, I'm reliving the terrible things that were done to me and other boys. I have a beautiful wife and two great kids who are approaching their teens. I thank God that the things that happened to me won't happen to them.

My girlfriend says that my sperm smells very strange inside her after we make love. At first I was hurt by her remark, but I must say that she's right. What causes this?

Mary Anna Friederich, M.D., an obstetrician-gynecologist, notes that semen (which includes sperm) and vaginal secretions have their own particular smells. When they're mixed together, they produce a special odor because of chemical interaction. Thus, the woman should expect a vaginal odor after intercourse. It will disappear in a few hours as the vagina cleanses itself.

My girlfriend had a baby, and she says it's mine. I'm not so sure about that, because I know she was also going with someone else.

I'm going to have a test to see if I'm the father. Are these tests accurate? Will I know for sure?

Blood tests to establish paternity are considered about 95 percent accurate.

One commonly used type of test is the human leukocyte group A (HLA) antigen test, an advanced form of tissue-typing that uses an analysis of genes. So if you're told that you're the father of the child after the results of this test are in, the chances are overwhelming that you are.

I'd like to know the figures on how many teenagers are having intercourse.

Statistics do not support the public image of rampant, promiscuous sexuality among young people. Several large studies suggest that:

• About 12 million of the 29 million young people in this country between the ages of 13 and 19 have had intercourse. This means that over half of all teenagers have not had intercourse by the age of 19.

• Of the 12 million who have had intercourse, about 7 million are males, and 5 million are females.

• The number of sexually experienced teenagers rises with age. Thus, while just one-fifth of 13- and 14-year-olds have had intercourse, nearly half of 15- to 17-year-old males and one-third of 15- to 17-year-old females have had intercourse at least once.

• Most teenagers who are not virgins have intercourse infrequently.

• Most teens have had only one or two sex partners.

Can you tell me something about giardiasis? My doctor says I have this disease and suggests that it could have been sexually transmitted.

Giardiasis is an intestinal infection caused by a one-celled protozoan called *Giardia lamblia.*

The disease is becoming increasingly common in the United States. Although it is not a reportable disease, the Centers for Disease Control recently identified over twelve thousand cases in a single year.

Giardiasis is transmitted through exposure to contaminated feces. This may occur by hand-to-mouth contact, by drinking contaminated water, or by eating raw fruits and vegetables that have been exposed to infected feces.

The infection can also be transmitted sexually, through ano-lingual contact or when oral-penile contact follows anal intercourse.

The most common symptom of giardiasis is a watery, foul-smelling diarrhea. Other symptoms include abdominal pain, loss of appetite, weight loss, nausea, and weakness.

Treatment is usually with the drug quinacrine hydrochloride (Atabrine) for five days. Side effects may include dizziness, headache, vomiting, and a yellow tinge to the skin. The extreme bitterness of the drug can be ameliorated by crushing the tablet and mixing it with applesauce.

Alternative treatment is with the drug Flagyl.

When I went to boarding school, I heard that saltpeter was added to the food in order to lower the male sex drive. I heard the same story about the army and prisons.

Do these institutions really put saltpeter in food? And does it lower sex drive?

No to both questions.

Saltpeter (potassium nitrate or sodium nitrate) is used mainly as a meat preservative. It does not affect sexual interest or capacity.

No responsible institution would add saltpeter to its food to suppress the sex drive. To do so would risk poisoning the diners since, in toxic quantities, saltpeter causes vomiting and severe gastro-intestinal distress.

In the eighteenth and nineteenth centuries, saltpeter was used to bring down fevers. It was erroneously believed to cool the body. Folk wisdom associates sexual excitement with heat, so the myth that saltpeter lowers sex drive may have been inspired by its supposed effect on fevers.

By the way, being in an all-male institution is a factor that can lower sex drive. In prisons, many men find that they have fewer erections than usual because of the lack of sexual stimulation.

Lately, I've been having emissions from my penis of, I think, semen. They happen while I'm going about my daily affairs—without my getting an erection, without an orgasm, and without any sexual stimulation whatsoever.

Usually, about half a teaspoon is emitted. Why is this happening? I'm 25 years old.

See your doctor.

Since the emission occurs without sexual stimulation or orgasm, it's unlikely to be a discharge of semen. It's far more likely that the discharge is pus from an infection. The infection could be caused by a sexually transmitted disease such as gonorrhea. Or it might be from an infection of the prostate or the urinary tract.

In any case, it requires a physician's treatment.

I enjoy having anal sex with my husband. His penis is not all that large. Yet, during sex and the following day, I have a lot of anal bleeding. And I have pain when I move my bowels.

Do you think I have a physical problem or should I use lubrication other than saliva?

The anal canal has very little natural lubrication. Thus, artificial lubrication—such as petroleum jelly or K-Y Jelly—is necessary to promote insertion without injury to delicate tissue. Saliva is not sufficient.

You need to see a physician. Bleeding and bowel movement pain after anal intercourse are often caused by tears (fissures) in the anus. Repeated irritation from anal friction can result in severe pain, bleeding, and a pus-laden discharge. Your doctor may recommend painkillers, stool softeners, and antibiotics if necessary. Condoms are recommended for anal intercourse.

Our 11-year-old son is small for his age. In addition, he has a moderate case of Tourette's disease, a disorder characterized by tics such as uncontrolled blinking, grimaces, shoulder shrugging, arm movements, and vocal tics—shouting or barking.

So he has had his share of trouble. What worries us is that he only wants to play with the girls at school. He hates any form of competition and gets teased by the boys.

Now he says he wants to dress up as a girl for Halloween. We're worried about this even though he appears to have a higher than normal heterosexual interest in women's breasts and sexy posters.

Are there any signs to watch for of homosexuality?

Two specialists in gender disorders, Tom Mazur, M.D., director of psychoendocrinology at the Children's Hospital of Buffalo, and Michele Youakim, a graduate assistant, comment that there are, as yet, no "signs" that signal homosexual development. For that matter, there are no absolute indicators of heterosexual development either.

As you know, many stereotypically masculine males are homosexual. And many men who appear "feminine" are exclusively heterosexual.

However, recent evidence suggests that some boys who prefer the company of girls, as well as girls' toys, clothes, and make-up, may grow up to have a homosexual or bisexual orientation. Not all these boys develop such orientations. Some apparently develop heterosexually. It's far from clear how we evolve our sense of gender identity and our sexual orientation.

You mentioned that your son has Tourette's disease. There is no evidence to link homosexuality with Tourette's. It may be, however, that your son avoids competition and plays with girls because they make him feel safer. You mentioned that the other boys tease him. He may find that girls are more accepting of his condition.

You should get a thorough evaluation of the effect that Tourette's disease has had on your son's psychosocial and psychosexual development.

My friend (we're 16-year-old girls) says that you won't get venereal disease if you have oral sex instead of intercourse.

Is this true?

This is a common misconception among teenagers—and among many adults as well.

In fact, a number of venereal diseases can be transmitted orally. Gonorrhea and syphilis germs can infect the throat. Thus, they can be passed on by oral sex. Viral diseases, such as herpes simplex and cytomegalovirus infections, can also be spread through oral contact. The presence of a cold sore or any sore on the mouth should rule out oral sex.

Pathogens from a mouth or throat infection—such as an abscessed tooth or strep throat—can be transmitted through cunnilingus. The result can be a vaginal inflammation or infection.

AIDS is also transmissible through oral sex.

I am a 31-year-old transvestite. What psychological effects will my cross-dressing have on my children? I have three daughters, ages five, four, and two.

I don't wear dresses around the house. But I do prefer to wear a nightgown to bed and feminine underwear under my own clothes.

You're wise to be concerned about the possible effects of your cross-dressing on your daughters' emotional health.

Even though ''transvestism may be harmless in that this behavior is not assaultive or abusive, it has other consequences from the point of view of children and their emotional development,'' says psychiatrist David W. Krueger, M.D., clinical associate professor of psychiatry at Baylor College of Medicine in Houston.

Dr. Krueger observes that children imitate and identify with both parents. When the father cross-dresses, his behavior becomes a model. In one case, the three sons of a transvestite father "identified strongly with him and his attributes, and became transvestites themselves."

We've found no research that specifically addresses the question of what effect your transvestism might have on your daughters. But children in their pre-school years are engaged in the task of sorting out the differences between the sexes and in establishing an identification with one or the other. The fact that you wear clothing that is strongly associated with females could be confusing and perhaps disturbing to your daughters.

You can avoid any potential problems by maintaining your privacy. Your daughters needn't know what sort of undergarments you wear. As for sleeping clothes, you might substitute a man's nightshirt for a woman's frilly nightgown.

Sex educator Judy Henkel suggests that you select nightshirts made of soft material—for example, nylon or the newer silklike polyesters. That way, if your daughters see you in the morning, you'll be dressed appropriately, but in a way that may still satisfy your need for cross-dressing.

I'm a 26-year-old man. I like to masturbate by putting a dildo or other object into my rectum. Is this a dangerous thing to do?

Yes. Foreign bodies inserted into the rectum—artificial penises (dildos), vibrators, cucumbers, candles—may cause severe tissue damage. The rectum joins the sigmoid colon at a right angle. Objects can thus move from the rectum into the colon, where they cannot be reached. In the colon, they may cause perforation and even fatal peritonitis.

If an object is lost in the rectum, no attempt should be made to retrieve it. The victim should go to a physician's office or a hospital emergency room. Anesthesia and special instruments may be required to remove the object.

■ *When they say some method of birth control is 90 percent effective, does it mean that 10 percent of the people using it get pregnant? Or does it mean that you can get pregnant ten out of a hundred times you have sex? Or 10 percent of the time in your lifetime? Or what?*

"Ninety percent effective" means that ninety out of a hundred fertile couples using the method for one year will not have an unplanned pregnancy.

Most figures reflect the effectiveness of contraceptive methods if they are used perfectly every single time. But since people do not use birth control perfectly every time, there is often a great gap between the potential and the actual effectiveness of various methods.

For example, the National Clearinghouse for Family Planning Information says that foam is 97 percent effective if used perfectly every time. But surveys of couples who use foam have found it to be only 84 percent effective.

Birth control pills are potentially the most effective contraceptives. The Pill with combined estrogen and progestogen is 99.66 percent effective if used perfectly every time, but 90 to 96 percent effective based on actual use.

The Mini-pill (with progestogen only) is 98.5 to 99 percent effective if used perfectly, 90 to 95 percent effective based on use.

The IUD is the next most effective contraceptive—97 to 99 percent effective if used perfectly. It's 95 percent effective based on actual use. The condom and the diaphragm are each 97 percent effective if used perfectly, and 90 percent effective based on use.

■ *Is it true that intercourse with uncircumcised males is a possible cause of cervical cancer in females?*

Studies do not confirm the widespread belief that circumcision prevents cancer of the cervix in the female partner—or that lack of circumcision could cause cervical cancer.

The myth developed because Jewish men are circumcised and Jewish women rarely develop cancer of the cervix (the narrow opening to the uterus at the top of the vagina). Researchers have now found that the supposed immunity of Jewish women to cervical cancer was probably due to a sexually conservative life-style rather than to male circumcision.

In general, cervical cancer seems to be associated with having early intercourse and many partners. The disease is rare among virgins. Early childbearing also seems to put a woman at greater risk of cervical cancer.

Exposure to herpes simplex virus type 2 (genital herpes) is also associated with cancer of the cervix, although how the two interact is not clear. But it has been found that women who suffer from cervical cancer show a very high incidence of antibodies to herpes, indicating that they have been exposed to this virus.

Cancer of the cervix can be detected by the Pap test, an inexpensive, painless procedure. If detected early, it is one of the most curable forms of cancer.

My girlfriend wants to have intercourse even while she's menstruating. Is it safe?

There's no health-related reason to abstain from intercourse during menstruation, observes Mary Anna Friederich, an obstetrician-gynecologist.

Despite widely held fears, menstrual flow is not "unclean." The flow contains an ounce or two of blood, plus cells and tissue fluid from the degenerating layers of the lining of the uterus. In no way is the flow harmful to either partner. Nor is it likely that penile thrusting will force menstrual fluid into the fallopian tubes, another common fear.

Some couples find intercourse especially enjoyable when the woman is menstruating. Many women experience an increase in sexual desire at the start of a period. The very moist vagina can enhance sensation, and the engorgement of the blood vessels in the woman's pelvis can heighten her orgasm. Orgasm, in turn, may relieve menstrual cramps, which are caused by congestion in the blood vessels.

Some couples are loath to have intercourse during menstruation because they think it will be messy. There are ways to handle this. If the woman regularly uses a diaphragm, she can insert it before intercourse. The diaphragm will catch the menstrual flow. It's also helpful to keep a towel or a supply of tissues at bedside so that the couple can clean up afterward.

I'm a 23-year-old graduate student. I'm frequently unable to get or keep an erection, but my doctor finds nothing wrong with me.

I smoke two packs of cigarettes a day. Could smoking be contributing to my impotence?

Indeed it could, warns Terrence R. Malloy, a urologist.

Smoking two packs of cigarettes—or less—a day can greatly constrict your blood vessels, thus limiting the blood supply you need for an erection. Smoking also impairs lung function. Inadequate breathing can tire you and cause you to lose your erection during active intercourse.

Moreover, smoking may leave odors on your breath and body. You may find your partners pulling back somewhat from you. Even if you're not conscious of being rejected, you may experience a diminishment of sexual arousal.

CONFESSIONAL BOX

"I don't want to jeopardize my marriage"

James, (31): I find it difficult to explain what I feel about being gay. I have a loving wife and two wonderful children, so it would be difficult to "come out." I don't think that I would want to jeopardize fourteen years of marriage, but I feel as though someone has tied me into a knot. I want to have a gay life, yet I'm very happy with my straight life. So far no one, to my knowledge, has an inkling of my preferences. I sure would love to tell someone of my feelings. It's not easy. I just wish I could have a gay-lover "mistress." It would sure alleviate a lot of pressure.

I've had affairs with a number of attractive single women. All of them knew that I was married and had no intention of leaving my wife and children.

Still, these women usually ended up feeling hurt—so why did they get involved with me in the first place? Frankly, I'm not all that handsome, charming, or great in bed.

Even if a married man isn't Mr. Wonderful, he may be attractive to a single woman for several reasons. Here are some of those suggested by psychiatrist J. Elizabeth Jeffress of the University of California School of Medicine, San Francisco:

• The single woman relishes the competition with another woman. The man's apparent unavailability for marriage only serves to increase the challenge.

• She thinks a married man is more mature and settled. She won't have to "train" him, the way she might a younger, less experienced man.

• She enjoys comforting a married man, as if he were a Daddy who came to his daughter for understanding.

• She has a deep need to restore the disrupted mother-father-daughter triangle of childhood.

• She's ambivalent about marriage and may prefer a married man because he's unavailable.

• She enjoys the special status of being a mistress.

• She considers the available single men to be "rejects."

Although it's commonly believed that the "other woman" suffers in an affair, psychiatrist Jeffress says that the man's wife is the person who really gets hurt.

My husband smokes marijuana, and since we'd like to have a baby soon, I'm worried. Could smoking pot cause him to become infertile?

To give yourselves the best chance of conceiving, your husband should discontinue the use of marijuana for at least six months before you attempt to become pregnant. That's the recommendation of Wylie C. Hembree, M.D., associate professor of clinical medicine and clinical obstetrics and gynecology at the Columbia University College of Physicians and Surgeons.

Your husband should give up marijuana immediately if tests have already shown that his sperm production is deficient.

Although marijuana use has not been conclusively linked with male infertility, several studies suggest there may be a connection. Research on both animals and humans has demonstrated that marijuana alters the process of sperm production and may decrease both sperm number and motility (movement).

In most men, says Dr. Hembree, this decrease in sperm production may not be great enough to cause infertility. But in men whose sperm production is already low, "depression of sperm production by marijuana could significantly reduce chances of conception."

A few months ago I sustained pelvic injuries in a motorcycle accident. Now I have what my doctor calls "retrograde ejaculation." I can have an orgasm, but no ejaculate comes out. Instead, it goes into my bladder.

I find this wonderful since my wife and I (I'm 27) no longer need birth control. But if she ever wants to become pregnant, I figure I can urinate into her vagina, since the sperm are in my urine. Am I right?

In retrograde ejaculation, as you have observed, semen is discharged into the bladder rather than out through the urethra. The condition can result from pelvic injury, as in your case, as well as from spinal cord injury, prostate surgery, diabetes, medications, and some types of genitourinary surgery.

Retrograde ejaculation does prevent conception, but it's not easy to create a pregnancy by urinating into a woman's vagina—sperm do not survive well in urine. Although a few cases of women getting pregnant that way have been documented in the medical literature, success is very rare, and nothing you can count on if you want a child. However, several techniques have been devised for collecting sperm from the urine and preserving it. Then it is used to artificially inseminate the woman. Even with this modern technique, it's difficult to achieve a pregnancy.

While my wife and I were making love, and I was reaching orgasm, the condom came off inside her.

Can my wife become pregnant? If so, is there any way to prevent it?

Yes, there is a possibility of your wife being pregnant. Even with careful use of the condom, there's a failure rate of about 3 percent. But when the condom slips off after ejaculation, semen spills into the vagina. It's virtually the same as having used no contraception at all.

In the future, if a condom fails for any reason, your wife may be able to avert pregnancy by immediately applying a spermicidal preparation, such as jelly, cream, or foam. The most effective thing to use after condom failure is spermicidal foam. It would be a good idea to keep it handy near your bed.

An even better method of protection is for your wife to use foam during your regular lovemaking while you wear a condom. Not only will she be fairly safe if the condom fails, but the combination is also one of the most reliable forms of birth control. The protection it provides is as good as the Pill—over 99 percent effective—but it doesn't have the Pill's side effects.

To prevent a condom from slipping off during intercourse, make sure that you remove your penis right after ejaculation, before it becomes soft. Hold the edge of the condom with your fingers to prevent it from slipping off as you remove your penis.

Move away from your wife as you take the condom off. If you remove the condom in close contact with her vaginal area, there's a possibility that sperm could migrate into the vagina and cause pregnancy.

My older sister says she's sworn off sex for a while. She hasn't slept with anyone for almost two years.

Is this normal? She's 33 and has never been married.

There are many valid reasons for someone to voluntarily refrain from sexual contact.

People often abstain from sex during periods of emotional stress.

Some may find themselves so preoccupied with work or creative activity that they don't think about sex for months or even years at a time. Others find that in the absence of a particular sex partner, they are not interested in sex at all.

Still others abstain from sex for a time following a series of unpleasant sexual relationships. They find a sexless period a comfortable respite from the dating game.

Many of those who give up sex for a time find the experience instructive and satisfying. They often have a sense of independence and freedom in not feeling compelled to engage in sexual activity simply because they have the opportunity.

Others, however, are troubled by their lack of interest in sex, particularly if a period of abstinence stretches into years. Such people may seek help from sex therapy clinics. Indeed, sex therapists find that a lack of sexual desire has become one of the most prevalent complaints in recent years.

Often, medical factors are found to account for diminished desire. These include illness, medication, and depression. Sometimes, neurotic conflicts may account for abstinence. In such cases, psychotherapy or sex therapy may be helpful.

If a person is comfortable with abstinence, it is not harmful. Lack of sex causes no organic illness or medical condition. However, people who repeatedly become sexually aroused without sexual release are likely to experience pelvic congestion and discomfort.

Nor does abstinence cause psychological disturbances. A popular notion holds that abstinence leads to frustration, which causes neurosis. There is no evidence for this belief.

My boyfriend and I (we're both in our twenties) have oral sex a lot, and I like to swallow his semen. Since I have a weight problem, I wonder about the calorie count of semen. Can it make me fatter?

You needn't worry that swallowing semen will significantly contribute to your weight problem.

One researcher has estimated that the average ejaculate, given its protein and fat content, yields less than fifty calories—about the same as a cracker or two.

My penis has fine hair growing on the shaft from the base up to the circumcision line.

When I have an erection, bumps appear at the roots of these hairs. It looks as if I have goose bumps or chicken skin. Not only do I feel embarrassed, sometimes the bumps become infected and a pimple appears.

Is there any way to cure this problem? Would it help to remove the hairs?

Hair may grow on the shaft of a penis because there is tight scrotal skin close to the circumcision line, according to Terrence R. Malloy, M.D., a urologist. The bumps that appear at the roots of these hairs may be sebaceous cysts.

Consult a urologist. He or she may be able to revise the circumcision and place skin that cannot produce hair in the area of the shaft of the penis.

What causes yeast infections?
I get one from time to time, but I don't understand why. I'm a 26-year-old woman.

Yeast infection (also called moniliasis, candidiasis, and fungus infection) is a very common vaginal infection. It's caused by the fungus *Candida albicans.*

The spores of this organism usually inhabit the vagina without causing any harm. But when the normal balance of bacteria, fungi, and other organisms in the vagina is disturbed, symptoms of yeast infection appear. These include genital itching, redness, and a thick vaginal discharge that resembles cottage cheese.

Women often develop yeast infections while taking antibiotics for infectious disease. Antibiotics kill the normal bacteria in the vagina, causing the yeast to overgrow. Pregnant women and those with diabetes are also more susceptible to yeast infections because the balance of their vaginal organisms has been changed.

Taking birth control pills may also promote yeast infection, since the pills lessen the acidity of the vagina. Vaginal douching, or bathing in bubble baths, can also alter the normal acid/alkaline balance of the vagina, resulting in yeast infection.

A woman is more susceptible to yeast infections when her resistance is low from another infection or illness. Similarly, emotional stress may contribute to yeast infection. So can poor diet, lack of sleep, or drug use.

Many women suffer from flare-ups during menstruation, when the acid/alkaline balance of the vagina changes.

The infection is sometimes transmitted through sexual intercourse. Many men who have yeast infections show no symptoms. But they can pass the infection back to their usual partner through intercourse. Thus, a woman who seems to have a chronic yeast infection may actually be constantly reinfected by her partner.

I was pregnant for six weeks without knowing it. Of course, I stopped taking the Pill as soon as I realized the situation. But will my baby be harmed by my having taken the Pill for six weeks?

There's no cause for concern.

Recent studies have shown that a woman who takes birth control pills before realizing she is pregnant does not increase her chances of having a child with birth defects. Nor is there a greater chance that she'll have a spontaneous abortion.

Many birth control pills contain the female hormone progesterone. In the sixties, it was feared that progesterone might adversely affect the development of fetal organs early in pregnancy. However, a large number of studies done in the past decade have not found a higher rate of birth defects in babies born to women who took progesterone.

I've heard that AIDS might be passed by mosquito bites. I know that other diseases are spread this way. So, is it true of AIDS?

There's no evidence that any cases of AIDS have been spread by mosquitoes or other insects. In fact, researchers at the Centers for Disease Control in Atlanta consider such a mode of transmission highly unlikely.

It's true that dozens of diseases—including malaria, encephalitis, and dengue fever—are known to be passed by mosquito. But the virus that causes AIDS (the HTLV-III virus) is extremely selective about the type of cell it infects.

Thus far, the virus is known to attack only a particular type of cell in humans and chimpanzees. It is not known to have any host relationship with mosquitoes. What's more, researchers feel that the pattern of AIDS infections suggests that AIDS is not transmitted by mosquitoes. If it were, it would be much more common.

My husband and I use foam as a contraceptive. We like to have oral sex before intercourse, but he doesn't like the taste of the foam. Is it alright if I apply the foam after intercourse?

At its best—that is, applied before intercourse—contraceptive foam alone is not a particularly effective form of birth control. It's been found to have a failure rate of about 15 percent, meaning that of 100 fertile couples using foam for a year, 15 will have an unplanned pregnancy.

Foam's effectiveness when used as contraception after intercourse is practically nil. Sperm can travel rapidly through the uterus and into the fallopian tubes, making contact with an egg within a minute or so after ejaculation. Thus, even if you insert foam immediately after intercourse, it may already be too late.

Foam should be inserted afterwards only in the case of condom failure—such as a leak.

My friend told me that drinking ginseng tea increases sexual desire and potency.
Is this true?

There's no evidence that ginseng has any effect on sexual desire and performance. It is a plant that interacts with the endocrine system in ways that are not completely understood.

Ginseng was first used as a supposed sexual stimulant in ancient

India. In contemporary oriental medicine, it's still considered a remedy for all kinds of sexual inadequacy. For example, it has been claimed to restore fertility. Among some American Indians, ginseng is thought to improve sexual desire, attractiveness, and performance.

A recent study of ginseng users in the United States found that only 7 percent felt it enhanced sexual performance. Many, however, reported a general feeling of well-being, reduced fatigue, and stimulation. As with other stimulants, such effects can sometimes reduce sexual anxiety and lower inhibitions. This may increase the enjoyment of sexual activity.

But claims of increased sex drive and restored fertility remain unproved.

I'm an inexperienced 22-year-old man. How can I tell when it's the right time to put my penis inside the woman?

Many people think of sexual intercourse as a silent dance that's been choreographed for them. They believe they must learn the right moves to make it all happen in the right way. It's much better to think of sexual activity as a mutual improvisation. You and your partner are working out your own variation on a theme.

Don't worry about "technique." Experts agree that the best technique is affection for the sex partner and the wish to give pleasure while receiving it.

Instead of concentrating on sexual skills, couples are best off concentrating on having fun, on being caring and involved. It's also best if you think of penetration—the penis entering the vagina—as just one aspect of a range of enjoyable sexual activity. If you consider penetration the only "real" sexual event, you're likely to have a lot more anxiety about your sexual performance.

In practical terms, penetration should occur after the woman has been stimulated enough for her vagina to lubricate, thus allowing the penis to glide in with ease. The amount and kind of sexual activity necessary to achieve this condition varies considerably from woman to woman.

How can you tell what's right for your partner? Ask! In words or gestures—or both—communicate your sexual preferences, and encourage her to communicate hers. Remember that lovers are not mind readers. Each partner must let the other know what is most pleasurable, without insisting on sexual activities the other does not enjoy.

Many couples evolve a signal for the time penetration should occur—a whispered "Now," for example, or a change of position that indicates readiness. Or the woman may simply guide the penis into her vagina when she feels ready.

In new relationships, if you're unsure of your partner's readiness, you might ask, "Now?" or "Ready?" Let her responses be your guide.

A lot of my friends use withdrawal for birth control. They say it works very well. I say it's not much good. Who's right?

You are.

While better than using no method of birth control at all, withdrawal is woefully subject to failure. The method is considered only about 75 to 80 percent effective. In contrast, the Pill is over 99 percent effective, if used correctly.

Withdrawal (coitus interruptus) is probably the oldest form of birth control. It requires the man to withdraw his penis from the woman's vagina just before he ejaculates.

In theory, this common method costs nothing and is always available. But the woman can get pregnant if the man is slow in withdrawing and deposits even a drop of semen in her vagina. Preseminal fluid may also leak into her vagina long before he ejaculates. It contains live, active sperm, which can cause pregnancy. Also, if he ejaculates near her vagina, some sperm may get inside and swim up into her uterus.

Withdrawal makes great demands on the man's self-control. The split-second timing required can interfere with a couple's enjoyment of intercourse. All in all, your friends would be well advised to seek out a more effective and emotionally comfortable method of contraception.

Is there something wrong with me? I'm an 18-year-old guy, and I find that I usually enjoy masturbating more than having sex with my girlfriend.
Help!

Research has shown that masturbation is likely to provide a more intense physiological response than intercourse. That's because an individual can concentrate entirely on personal sensations, without having to think about a partner's needs. One can let go and self-centeredly enjoy sexual feelings.

In addition, intercourse with your girlfriend is likely to take place under less than ideal circumstances. If you both live at home, your sexual encounters may have to be secretive and fleeting. You may also be suffering from performance anxiety—wondering how you're doing, if your responses are normal, if you are satisfying your partner.

Another anxiety that interferes with sexual enjoyment is fear of pregnancy and of sexually transmitted diseases.

Any of these factors may account for the fact that you find masturbation more enjoyable than intercourse.

Here's the problem: My boyfriend wants me to do oral sex on him, and I'd very much like to. But I'm scared that I'll gag and choke.
Also, what are the rules for oral sex? Should couples take turns—or do it on one partner at a time? What else should I know to do it right?

Richard Reznichek, M.D., and his wife, Carol Reznichek, R.N., a sex therapy team, say it's best to make oral sex playful. A good way to start is to put flavors such as maple syrup, whipped cream, or chocolate sauce on your partner's genitals and lick them off.

Since you have a fear of gagging, you might want to begin by just licking the shaft of the penis, or sucking on only the head of the penis. Make sure that you're in a position where you can quickly disengage your head if you feel a choking sensation.

There are no rules for "correct" oral sex. It can be done simultaneously or serially. Your best guide for what's correct in your own relationship is good communication with your partner about what you both find pleasurable.

Often men find the frenulum—the fold on the lower surface of the head of the penis—particularly sensitive to oral stimulation. Rhythmic sucking of the penis gives pleasure to many men. Some find the licking or sucking of the testicles or perineal area gratifying.

Over the past five years, my wife's sex drive has decreased drastically. Five years ago, we had sex three to five times a week. Then it diminished to twice a week. Now, we make love once a week—if I'm lucky.

My wife seems to have lost most of her sex drive. Could she be sick? I'm 36 and she's 34.

A great many factors may account for—or contribute to—a decline in sex drive.

A physical condition is one possibility. Sometimes decreased sexual interest is an early symptom of diabetes or multiple sclerosis. But while physical ailments should be considered, it's much more likely that other factors are responsible.

Just plain tiredness may contribute to a woman's lack of interest in sex. That's especially true if she's trying to juggle a career, household chores, child care, and marriage. Stress and anxiety may also contribute to a decline in libido. So may conflicts in your wife's relationship with you. If fatigue is the problem, try getting some household help or sharing more child-care and housekeeping responsibilities with your wife.

Another reason for your wife's sexual disinterest may be clinical depression, often overlooked as a cause of decreased sexual drive. Since people don't always know when they're depressed, look for such symptoms as insomnia or irregular sleeping patterns. Diminished appetite and weight loss are also common symptoms of depression, as are fatigue and agitation. If your wife is depressed, she may also suffer from headaches, backaches, and hypochondria.

Is your wife on any medications? If so, she may be suffering sexual effects. For example, medications that control high blood pressure may decrease sexual desire in women as well as men. Oral contraceptives and tranquilizers, such as Valium, can also cause a decline in sexual interest.

While you're looking for reasons, however, don't overlook the possibility that your sex life may have grown boring. Simply perking it up by changing your routine and learning new techniques may revive your wife's flagging interest in sex.

I'm a 17-year-old virgin. How many other girls my age have already had intercourse? Has the percentage increased in the past few years? From the talk I hear, I'd say it has.

A 1979 study of white girls at Johns Hopkins University found that 53 percent had experienced intercourse before their eighteenth birthdays.

A 1984 study reports virtually the same results. In a survey of sexual activity among teenagers at two suburban high schools near Chicago, investigators found that 56 percent of the 17-year-old girls had had intercourse. The researchers believe that the parity of figures between 1979 and 1984 indicates that the sexual revolution may have leveled off. Middle-class adolescents, they suggest, may be "trying to strike a balance between a more permissive sexual ideology, their parents, and, in recent years, society's increasingly more conservative sexual attitudes."

Another speculation is that as the proportion of adolescents in the United States declines, there is less peer pressure on adolescents to engage in early sexual activity.

In general, the researchers found, youngsters who performed at an above-average level in school were less likely to have had intercourse than students who were average or below average. Students who lived with both natural parents were also less likely to have engaged in intercourse.

The teenagers surveyed in Chicago revealed a great deal of ignorance about venereal disease and birth control. Nearly half the boys admitted that they knew "little or hardly anything" about sexually transmitted disease.

Myths about pregnancy were common: Many of the students

erroneously believed that a girl cannot become pregnant the first time she has intercourse.

There was a discrepancy between knowledge and behavior in regard to birth control. Although two out of three students claimed to know "a lot or quite a bit" about the subject, relatively few sexually active teenagers used contraception consistently and reliably. In fact, less than 10 percent of the sexually active teenagers always used some form of birth control.

Fully 20 percent said that they and their partners never used birth control. Not surprisingly, 12 percent of the sexually active girls in the study reported having been pregnant.

I've had transsexual tendencies all of my life (I'm 29 now). This is the first time I've ever mentioned them to anyone.

I travel a lot for my work, and when I do, I cross-dress and go out as a female. I spend all my money on items of women's clothing and can't wait to get out of town to wear them. I'm married and have two children. My wife knows nothing about my cross-dressing. I'm sure that she would leave me if she did.

I'd like to get some relief from pressure by at least talking with others who have the same problem. Can you help?

We referred this question to Roger E. Peo, a sex educator who received his doctorate from the Institute for Advanced Study of Human Sexuality.

Dr. Peo points out that many transvestites call themselves transsexual because they enjoy cross-dressing so much. In fact, transvestites do not have the transsexual's desire to be a woman. A transvestite is a male who enjoys dressing in women's clothes and often gets sexually aroused from doing this. In contrast, transsexuals generally do not become aroused from wearing the clothes of the other sex.

Your preoccupation with cross-dressing probably stems from the fact that you've had to be so secretive about your activities. After all, you've never talked to anyone about your feelings, and even your wife does not suspect them. This undoubtedly adds to the pressures you feel.

Fortunately, your travel allows you some freedom without fear of discovery. Probably, if you had more freedom to cross-dress when you felt the desire to do so, you would be less preoccupied with cross-dressing.

Undoubtedly, your secret is affecting your relationship with your wife, even though you say she suspects nothing of your cross-dressing. However, since you can hardly wait to get out of town, she may suspect an affair with another woman. In any case, you're wise in seeking help in the form of contact with others. That will relieve your sense of isolation.

One thing you could do is to subscribe to *TV-TS Tapestry* magazine. It carries many articles dealing with transgender situations. The magazine has a section for people who wish to correspond with each other on transgender issues. If you get a post office box, you can receive mail in complete privacy.

In addition, the magazine publishes a listing of organizations that help people with concerns like yours. The magazine can be purchased for $10 from the Tiffany Club, P.O. Box 19, Wayland, MA 01778.

Another resource that may be of help to you is the Gender Identity Center of Colorado, Inc., 3715 32nd Avenue, Denver, CO 80211. The Center's 24-hour hotline number is (303) 458-5378.

I've been using BHT for two years to suppress herpes. The results have been excellent. Why don't you recommend it to your subscribers?

BHT is butylated hydroxytoluene. It's normally used as a food preservative, and it can be purchased in any natural food store.

Nicholas J. Fiumara, M.D., a venereologist, points out that the drug was investigated by the FDA and found to be without benefit. In your case, though, it has apparently worked well as a placebo. A placebo is a chemically valueless pill that relieves symptoms because the patient expects that it will. How placebos work physiologically is not understood, but placebos—no matter what their type—are effective in about 35 percent of cases.

Yes. When seven female sex partners of AIDS patients were examined, one was found to have the full-blown AIDS syndrome—including weight loss, swollen lymph glands throughout the body, a resistant type of pneumonia, and immunologic abnormalities.

Five other women studied had early signs of the disease. AIDS is caused by a virus and transmitted through body fluids such as blood and semen. The disease, from which no one has yet recovered, is most common among homosexual men and intravenous drug users.

According to the research on the women, which was reported in the *New England Journal of Medicine*, the only common risk factor among them was prolonged monogamous contact with an AIDS patient, lasting two to seventeen years. One woman showed early symptoms within eight months of beginning her relationship.

Anyone whose sexual partner is diagnosed as having AIDS is advised to discontinue sexual activity with the partner. He or she should be checked regularly for such early signs of AIDS as swollen lymph glands and a decrease in white blood cells.

I seem to have a new type of vaginal infection called gardnerella. Can you tell me more about it?

Gardnerella isn't a new type of infection—it's an old infection with a new name.

Gardnerella is caused by the bacterium *Hemophilus vaginalis*, and the infection used to be called hemophilus. Gardnerella is one of the most common types of vaginal infections. The *Hemophilus* bacterium is now known to be responsible for more than 90 percent of the cases that were formerly called "nonspecific bacterial vaginitis."

Gardnerella is characterized by a creamy-white or grayish discharge. There is also a foul odor to the discharge, often described as "fishy." The odor may be especially noticeable after intercourse. There may also be vaginal burning and itching.

The infection is usually treated with an antibiotic such as ampicillin.

Gardnerella can be transmitted sexually, so a regular sex partner should be treated at the same time as the woman. In one study, 79 percent of the male sex partners of infected women were found to be harboring the bacteria in their urethras. Without simultaneous treatment, the men would have been able to reinfect their partners after treatment.

In addition to being transmitted sexually, gardnerella may also occur when the normal acid/alkaline balance of the vagina is altered by other infection, stress, or hormonal changes.

Other instruments of transmission are thought to be douche nozzles, toilet seats, washcloths, and towels.

I masturbated quite often during my teens. Now I worry that perhaps my sperm supply was lessened as a result. Would abstinence improve the situation, or am I worrying needlessly?

I'm 30 years old, and I'd like to have children someday.

You're worrying needlessly. But you're expressing a common concern.

Many males mistakenly believe they have only a given supply of sperm. They fear that if they use up too much of it in adolescent masturbation, they won't have enough when they get older and want to have children.

In fact, in a healthy male the testicles will simply keep on producing sperm well into old age. When a man reaches his seventies, there may be a slight drop in the quantity.

I've heard that having sex on LSD is really great. Is that true?

Hallucinogens such as LSD (lysergic acid diethylamide) and mescaline are sometimes lauded as enhancers of sexual activity. Some users report an intensified sexual experience and heightened sensory awareness. Both men and women sometimes report "explosive" orgasm.

However, in interviews at the Masters and Johnson Institute with 85 men and 55 women who had used LSD on three or more occasions, fewer than 15 percent of each sex claimed that LSD enhanced sexuality. One researcher argues that taking LSD as an aphrodisiac is useless "because the user can't remain focused on what he started to do."

The variability in sexual response may be traced to the quality of the hallucinogens sold on the street. Often, LSD and mescaline have numerous other drugs and impurities combined with them. Then, too, the effects of drugs vary from person to person.

Although little scientific research has been done on the subject, in male rats, small doses of hallucinogens have been found to accelerate sexual behavior, while larger doses completely disrupt sexual activity.

I'm 23 and my girlfriend is 20.

We have an active sex life but lately we've also been doing some strange things. For example, sometimes we'll masturbate while talking on the telephone, while one of us describes a sexual fantasy. Also, I sometimes enjoy watching her masturbate.

How can you tell if what you're doing is normal? I'm starting to get worried about myself.

Sexual normalcy is a very tricky concept. Many sex therapists believe that any sexual activity between consenting adults that harms neither partner should be considered normal.

The activities you describe are certainly normal by this standard. What's more, they are common. Many people find mutual fantasizing and watching a partner masturbate sexually stimulating.

Here are some guidelines for deciding which sexual activities are normal for you: Are you hurting yourself or your partner, emotionally or physically? Is your behavior increasing—or decreasing—your sense of self-esteem? Are your sexual acts joyous, or are they producing anxiety and guilt? Do your sexual activities make you feel depersonalized, or are they contributing to your development as a whole person?

You can't measure your sexual behavior according to a standard of "normalcy," but you can determine whether your sexual behavior is bringing you pain or pleasure, and act accordingly.

CONFESSIONAL BOX:

"What's it all about?"

<u>Neil, (49)</u>: Sometimes I wonder what these messages on the service are really all about. Loneliness—certainly. There's also that strange American concern with virginity, along with the usual fetishists who delight in describing their own addictions. Lots of those innumerable personals could be shortened easily to read: "Man wants trouble-free sex for ego reasons," or "Woman looking" "Couples looking"—looking and looking for thrills and spills and who knows what else.

When you're a teenager, you think that there are answers, solutions to the confusion of your feelings. You think that at some point the movie music will reach a crescendo, the credits will roll, and everything will be all right. Then you grow up, and you learn what a broken thing being an adult can be: the nights when you can't sleep; the sense of loneliness when you're done with making love. You're left with this—a wish to reach out in a world of isolation and say something more than the ordinary garbage, to talk of the forbidden, and to listen quietly for an answer.

Human Sexuality
Consulting Editors

ARMANDO DeMOYA, M.D., and DOROTHY DeMOYA, R.N., M.S.N., were the first sex-therapy team trained by Masters and Johnson in a pilot training program for professionals.

Armando DeMoya is a board-certified obstetrician-gynecologist and a fellow of the American College of Obstetrics and Gynecology. He is on the faculty of Georgetown University Hospital and George Washington University Hospital.

Dorothy DeMoya has taught human sexuality at Georgetown University Medical School and is on the faculty and in the doctoral program at Catholic University School of Nursing. She was secretary of the Society of Sex Therapy and Research.

The DeMoyas write the ''Sex Q&A'' column for the nursing journal *RN*. With Martha and Howard Lewis they are co-authors of *Sex and Health: A Practical Guide to Sexual Medicine.*

They are co-directors of Washington Reproductive Associates in Washington, DC.

NICHOLAS J. FIUMARA, M.D., M.P.H., was director of the division of communicable and venereal diseases of the Massachusetts Department of Public Health.

His extensive teaching career has included positions as clinical professor of dermatology and preventive medicine at Boston University School of Medicine, clinical professor of dermatology and syphilology at Tufts University School of Medicine, and instructor in dermatology at Harvard Medical School.

He is the author of books and over two hundred medical journal articles, many on the subject of venereal disease.

MARY ANNA FRIEDERICH, M.D., is a board-certified obstetrician-gynecologist.

She is clinical associate professor of obstetrics/gynecology/psychiatry at the University of Rochester School of Medicine and Dentistry and has been an instructor and fellow in endocrinology at Duke University.

She is a member of the board of directors of the Society for Sex Therapy and Research. She has published articles and book chapters on the emotional and physiological aspects of pregnancy, menstrual problems, menopause, and chronic pelvic pain.

RICHARD V. LEE, M.D., is a board-certified internist with a special interest in adolescent medicine and sexually transmitted diseases.

Dr. Lee is a professor of medicine and chief of the division of maternal and adolescent medicine at the State University of New York at Buffalo. He was formerly on the faculty of the Yale School of Medicine.

He is head of the department of medicine at Children's Hospital of Buffalo, and is a consulting editor of the *American Journal of Medicine* and an editorial consultant for Viking Press and Year Book Medical Publishers.

He has published extensively in the medical literature. Some of his recent articles have concerned drug and alcohol abuse in pregnancy, toxic shock syndrome, adolescent sexuality, and the so-called new morality.

HAROLD I. LIEF, M.D., is a board-certified psychiatrist.

Dr. Lief is a professor of psychiatry and director of the division of family studies at the University of Pennsylvania School of Medicine.

He has been president of the Sex Information and Education Council of the United States (SIECUS), and is a recipient of both the Award for Outstanding Contributions to the Field of Human

Sexuality, presented by the Society for the Scientific Study of Sex, and the Annual Award of the American Association of Sex Educators, Counselors and Therapists. He has served on the faculty of the Center for Advanced Study in the Behavioral Sciences at Stanford University.

Dr. Lief has been a member of the editorial boards of such journals as *Medical Aspects of Human Sexuality, Journal of Sex Education and Counseling, Journal of Sex and Marital Therapy, Sexuality and Disability, Sexual Behavior, Archives of General Psychiatry, Journal of Divorce and Marriage*, and *Family Review*. He has published over two hundred articles and book chapters in the field of human sexuality.

TERRENCE R. MALLOY, M.D., is a board-certified urologist.

Dr. Malloy is clinical professor of urology at the University of Pennsylvania School of Medicine. He is chief of the urology section at Pennsylvania Hospital.

He has published widely in the medical literature. His recent work has included articles on cancer of the penis, penile prostheses, organic impotence, and transsexuality.

STUART MARGULIES, Ph.D., is a New York City consulting psychologist.

He is currently director of research at Mind Games, Inc. He is the author of seventeen texts and a wide variety of professional papers.

Dr. Margulies has extensively investigated the subject of "listening," identifying the barriers to effective listening and developing methods for improving listening. He has written several papers on how to listen in order to hear the hidden feelings behind the thoughts.

Among his other studies are an investigation of interpersonal difficulties people experience at work and of the difficult situations which must be resolved to diet successfully. He has also developed a widely used peer counseling system.

ARNOLD MELMAN, M.D., a board-certified urologist, is Physician-in-Charge of the Center for Male Sexual Dysfunction at Beth Israel Medical Center in New York City.

He is a professor in the department of urology at Mount Sinai School of Medicine in New York and an associate director of the department of urology at Beth Israel Medical Center.

Dr. Melman is editor of the journal *Sexuality and Disability* and has published widely in the field of male sexual dysfunction. He has presented scores of professional papers in this country and abroad.

He served as the program chairman of diabetes and impotence at the World Impotence Seminar in Paris in 1984 and was an invited panelist at the Fifth International Congress of Sexology in Washington, DC, in 1983.

ROGER E. PEO, Ph.D., has a degree from the Institute for Advanced Study of Human Sexuality.

His special areas of expertise are transvestism and transsexuality, and he conducts research and provides counseling for transvestites and transsexuals and their partners.

He is a professional advisor and writer for *TV–TS Tapestry*, a national magazine for cross-dressers and transsexuals, and is on the board of advisors of Metamorphosis Medical Research Foundation, a support organization for female-to-male transsexuals.

RICHARD C. REZNICHEK, M.D., and **CAROL G. REZNICHEK, R.N., M.S.N.,** are directors of the South Bay Center for Sexual Dysfunction in California. Both are certified sex therapists.

Dr. Reznichek is a board-certified urologist. He is clinical instructor in surgery/urology at the University of California School of Medicine in Los Angeles and is urologic consultant at the UCLA Human Sexuality Program.

He was chairman of the California Medical Association Venereal Disease Task Force and has served as a urologic consultant in medical schools and hospitals in Vietnam, Cambodia, Guatemala, and Malaysia.

Carol Reznichek has been an assistant professor of nursing at California State University in Los Angeles.

She was a consultant in maternal and child health in the department of obstetrics and gynecology at the Medical College of Georgia in Augusta.

She has been a consultant on family planning at the Torrance

Adult School and the Indochinese Refugee Center in Hawthorne, California.

BARBARA RISSMAN, M.S.W., is a certified social worker with an expertise in family therapy and adolescent problems.

She worked as a staff psychological counselor for students at Bard College in New York, and was a consultant family therapist for families of emotionally disturbed children at the Children's Annex in Woodstock, New York.

She currently works as a family therapist at the Maverick Family Health Center in Woodstock, New York.

RANDALL RISSMAN, M.D., is a board-certified family physician.

Dr. Rissman has taught family medicine at New York Medical College. He is director of the Maverick Family Health Center. He has a special interest in family-oriented psychotherapy and is medical director of the Catskill Family Institute in Kingston, New York.

The Rissmans are the authors of the forthcoming book *Links: The Family/Health Connection.*

SIDNEY ROSEN, M.D., a board-certified psychiatrist in full-time private practice in New York City, is the founding president of the New York Milton H. Erickson Society for Psychotherapy and Hypnosis, Inc.

He is assistant clinical professor of psychiatry at New York University College of Medicine, and author of *My Voice Will Go With You: The Teaching Tales of Milton H. Erickson, M.D.* (Norton, 1982).

Dr. Rosen served for twenty years as Psychiatrist-In-Charge of Psychiatric Services at the Rusk Institute of Rehabilitation Medicine at New York University Medical Center.

ESTELLE ROSEN, C.S.W., is a social worker with an expertise in marital and family therapy.

She is currently director of family therapy of the Bronx region for the Jewish Board of Family and Children Services. She also has a private clinical practice for marriage and family therapy.

She is adjunct professor in the Post Masters Program at New York University School of Social Work.

FRED WESTENDARP, M.D., a specialist in emergency medicine and a member of the American College of Emergency Physicians, is an expert on the health problems of gays.

He is active in the Phoenix area Gay Physicians Group and the Arizona Lesbian and Gay Task Force.

Dr. Westendarp has lectured on homosexuality to psychology and counseling students at Northern Arizona University, Arizona State University, and Phoenix College.

Index